XIX International Scientific Conference "RESHETNEV READINGS" 2015

IOP Conference Series: Materials Science and Engineering
Volume 122

Krasnoyarsk, Russia
10 - 14 November 2015

ISBN: 978-1-5108-2324-2
ISSN: 1757-8981

Printed from e-media with permission by:

Curran Associates, Inc.
57 Morehouse Lane
Red Hook, NY 12571

proceedings.com

Some format issues inherent in the e-media version may also appear in this print version.

Printed by Curran Associates, Inc. (2016)

For permission requests, please contact the Institute of Physics
at the address below.

Institute of Physics
Dirac House, Temple Back
Bristol BS1 6BE UK

Phone: 44 1 17 929 7481
Fax: 44 1 17 920 0979

techtracking@iop.org

Additional copies of this publication are available from:

Curran Associates, Inc.
57 Morehouse Lane
Red Hook, NY 12571 USA
Phone: 845-758-0400
Fax: 845-758-2633
Email: curran@proceedings.com
Web: www.proceedings.com

Table of contents

Volume 122

XIX International Scientific Conference Reshetnev Readings 2015
10–14 November 2015, Krasnoyarsk, Russia

Accepted papers received: 11 March 2016
Published online: 22 April 2016

IOP Conf. Series: Materials Science and Engineering **122** (2016) 011001 doi:10.1088/1757-899X/122/1/011001

XIX International Scientific Conference "RESHETNEV READINGS"

The International Scientific Conference "RESHETNEV READINGS" is dedicated to the memory of Mikhail Reshetnev, an outstanding scientist, chief-constructor of space-rocket systems and communication satellites.

The current volume represents selected proceedings of the XIX International Scientific Conference "RESHETNEV READINGS" held on November 10 – 14, 2015.

Plenary sessions, round tables and forums are regularly attended by famous scientists, developers and designers representing the space technology sector, as well as professionals and experts in the IT industry. A number of outstanding academic figures expressed their interest in the event including Jaures Alferov, Vice-president of the Russian Academy of Sciences (RAS), Academician of RAS and Nobel laureate, Dirk Bochar, General Secretary of the European Federation of National Engineering Associates (FEANI), Prof. Yuri Gulyaev, Academician of RAS, Member of the Presidium of RAS, President of the International Union of Scientific and Engineering Associations and Director of the Institute of Radio Engineering and Electronics of RAS, as well as rectors of the largest universities in Russia, chief executives of well-known research enterprises and representatives of big businesses.

We would like to thank our regular sponsors such as JSC "Reshetnev Information Satellite Systems", JSC "Krasnoyarsk Engineering Plant", Central Design Bureau "Geophysics", Krasnoyarsk Region Authorities. These enterprises and companies are leading ones in the aerospace branch. It is a great pleasure to cooperate and train specialists for them.

Peer review statement

All papers published in this volume of *IOP Conference Series: Materials Science and Engineering* have been peer reviewed through processes administered by the proceedings Editors. Reviews were conducted by expert referees to the professional and scientific standards expected of a proceedings journal published by IOP Publishing.

IOP Conf. Series: Materials Science and Engineering **122** (2016) 012001 doi:10.1088/1757-899X/122/1/012001

Combinatorial optimization in foundry practice

A N Antamoshkin[1,2] and I S Masich[1,3]
[1] Department of System Analysis and Operation Research, Siberian State Aerospace University named after academician M. F. Reshetnev, Krasnoyarsk, Russia
[2] Department of Economics and Information Technology of Management, Siberian Federal University, Krasnoyarsk, Russia
[3] Department of Informatics and Mathematical Modeling, Krasnoyarsk State Agrarian University, Krasnoyarsk, Russia

E-mail: i-masich@yandex.ru

Abstract. The multicriteria mathematical model of foundry production capacity planning is suggested in the paper. The model is produced in terms of pseudo-Boolean optimization theory. Different search optimization methods were used to solve the obtained problem.

1. Introduction

1.1. Production Capacity Optimization

One of the actual problems in modern industry is production capacity optimization under irregular orders from numerous partners. Along with mass serial production this orders may have small serial or single nature. The most orders are irregular, i.e. they cannot be planned beforehand but nevertheless they are enough profitable for the enterprise. It requires solving production scheduling problems many times in casual points of time.

So it is necessary to have means for including new small orders in capacity intervals of existing mass serial ones. Moreover, it is necessary to consider time of equipment revamping for small serial order since it take significantly more part of execution time then in mass serial production.

Existing of small serial and single orders requires practically permanent process of production capacity planning. Construction of production capacity program for industrial enterprise and its subdivisions is laborious and logically intricate problem. Consider it by the example of foundry practice.

Below we design an optimization model for production capacity and apply combinatorial methods to solve it.

1.2. Pseudo-Boolean optimization problems

Unconstrained pseudo-Boolean optimization is an issue that studied enough by now. Algorithms that have been designed and investigated in the area of unconstrained pseudo-Boolean optimization are applied successfully for solving various problems. Particularly, these are local optimization methods [1, 2] and stochastic and regular algorithms based on local search for special function classes [3-6]. Moreover, there are a number of algorithms for optimization of functions given in explicit form: Hammer's basic algorithm that was introduced in [7] and simplified in [8]; algorithms for optimization of quadratic functions [9-11], etc. Universal optimization methods are also used successfully: genetic algorithms, simulated annealing, taboo search [12, 13].

If there are constraints on the binary variables, one of ways to take into account it as is well known is construction and optimization of a generalized penalty function. Shortcoming of this approach is existence of a large number of local optima of the generalized function what will be shown below. If an accessible region is a connected one then this issue can be partly eliminated, for example, by using local search with a stronger system of neighborhoods. Extension of search neighborhood reduces the number of local optima which locate mainly not far one from another in this case.

If an accessible region is unconnected then using penalty functions and unconstrained optimization methods get complicated because the accessible region is usually too small with respect to optimization space. That makes difficult searching feasible solution.

In this work heuristic procedures of boundary point search are considered for a constrained pseudo-Boolean optimization problem. Experimental investigation of the algorithms are described, recommendations for their applying are given.

2. Search Algorithms for Pseudo-Boolean Optimization

2.1. Basic Definitions

Consider some definitions that are necessary for describing optimization algorithm work [14].

- A *pseudo-Boolean function* is called a real function of binary variables: $f: B_2^n \to R^1$, where $B_2 = \{0,1\}$, $B_2^n = B_2 \times B_2 \times \cdots \times B_2$.
- Points $X^1, X^2 \in B_2^n$ are called *k-neighboring* points if they differ in k coordinate value, $k = \overline{1,n}$. 1-neighboring points are called simply neighboring.
- The set $O_k(X)$, $k = \overline{0,n}$, of all point of B_2^n, that are k-neighboring to a point X, is called a *k-th level* of the point X.
- A point set $W(X^0, X^l) = \{X^0, X^1, \ldots, X^l\} \subset B_2^n$ is called a *path* between points X^0 and X^l if for $\forall\, i = 1, \ldots, l$ the point X^i is a neighboring to X^{i-1}.
- A set $A \subset B_2^n$ is called a *connected set* if for $\forall X^0, X^l \in A$ the path $W(X^0, X^l) \subset A$ exists.
- A point $X^* \in B_2^n$, for which $f(X^*) < f(X), \forall X \in O_1(X^*)$, is called a *local minimum* of pseudo-Boolean function *f*.
- A pseudo-Boolean function that has an unique local minimum is called an *unimodal* on B_2^n function.
- An unimodal function f is called *monotonic* on B_2^n if $\forall X^k \in O_k(X^*), k = \overline{1,n}$: $f(X^{k-1}) \le f(X^k)$, $\forall X^{k-1} \in O_{k-1}(X^*) \cap O_1(X^k)$.

Example: A polynomial of binary variables

$$\varphi(x_1, \ldots, x_n) = \sum_{j=1}^{m} a_j \prod_{i \in \beta_j} x_i ,$$

where $\beta_j \subset \{1,...,n\}$, is an unimodal and monotonic pseudo-Boolean function for $a_j \geq 0$, $j = \overline{1,m}$.

2.2. *Problem Statement*

Consider the problem of the following form

$$C(X) \to \max_{X \in S \subset B_2^n}, \tag{1}$$

where $C(X)$ is a monotonically increasing from X^0 pseudo-Boolean function,

$S \subset B_2^n$ is a certain subspace of the binary variable space; it is determined by a given constraint system, for example:

$$A_j(X) \leq H_j,\ j = \overline{1,m}. \tag{2}$$

In the general case a set of feasible solutions *S* is an unconnected set.

2.3. *Properties of a Feasible Set*

• A point $Y \in \mathbf{A}$ is a *boundary point* of the set $\mathbf{A}$ if there exist $X \in O_1(Y)$ for which $X \notin \mathbf{A}$.

• A point $Y \in O_i(X^0) \cap \mathbf{A}$ is called a *limiting point* of the set $\mathbf{A}$ with the basic point $X^0 \in \mathbf{A}$ if for $\forall X \in O_1(Y) \bigcap O_{i+1}(X^0)$ $X \notin \mathbf{A}$ holds.

• A constraint that determine a subspace of the binary variable space is called *active* if the optimal solution of the conditional problem do not coincide with the optimal solution of the appropriate problem without taking the constraint into account.

Consider some properties of a feasible set [15].

• If the object function is a monotonic unimodal function and the constraint is active then the optimal solution of the problem (1) is a point that belongs to the subset of limiting points of the feasible set *S* with the basic point X^0 in which the object function takes the lowest value:

$$C(X^0) = \min_{X \in B_2^n} C(X).$$

• Consider a problem (1) with a constraint (2). If the constraint function (2) is an unimodal function then the set *S* of feasible solutions of the problem (1) is a connected set.

A number of limiting points of a connected feasible set for the problem (1) equals $s \leq s_{\max} = C_n^{[n/2]}$, where $[n/2]$ is the integer part of the value $n/2$. In case of $s = s_{\max}$ all limiting point belong to $O_{n/2}(X^0)$ if *n* is even and to $O_{(n-1)/2}(X^0)$ (or $O_{(n+1)/2}(X^0)$) if *n* is odd.

2.4. *Transfer to Unconstrained Optimization*

One of ways to take into account constraints in conditional problems is construction a generalized penalty function.

Consider the generalized function

$$F(X) = C(X) - r \cdot \sum_{j=1}^{m} \max\{0, A_j(X) - H_j\} \tag{3}$$

for a problem (1) with constraints (2).

In this case the following lemma occurs [15]:

The limiting points of the set S of the feasible points of the primary problem (1)-(2) with the basic point X^0 are point of local maxima for generalized penalty function (3) with parameter r satisfying the condition

$$r > \frac{C(Z) - C(X')}{\sum_{j=1}^{m} \max\{0, A_j(Z) - H_j\}}$$

for every limiting point $X' \in O_k(X^0)$ and every point $Z \in O_{k+1}(X^0) \cap O_1(X')$.

So, a problem (1)-(2) is identical to the problem

$$F(X) \to \max_{X \in B_2^n}, \quad (4)$$

in which a number of local maxima (in case of connectivity of the feasible set) $s \le s_{\max} = C_n^{[n/2]}$. In general case, when feasible set is not connected, a number of local maxima theoretically can amount to 2^{n-1}.

The problem (4) is solved by known search methods: local search, genetic algorithms, simulated annealing, etc.

The essential shortcoming of this approach is loss of the monotonicity property for an object function $C(X)$. By addition of even simple (for example, linear) constraints the generalized function becomes a polymodal nonmonotonic function with exponential number of local maxima.

2.5. *Heuristic Algorithms for Boundary Point Search*

For any heuristic of boundery point search we will consider a pair of algorithms – primary and dual [16]. A primary algorithm starts search from the feasible area and moves in a path of increasing of the objective function until it finds a limiting point of feasible area. Otherwise, a dual algorithm keeps search in the unfeasible area in a path of decreasing of the objective function until it finds some available solution.

Total scheme of primary search algorithm

1. Put $X_1 = X^0$, $i = 1$.
2. In accordance with a rule we choose $X_{i+1} \in O_i(X^0) \cap O_1(X_i) \cap S$. If there are no such point then go to step 3; else $i = i + 1$ and repeat the step.
3. $X_{opt} = X_{i+1}$.

Total scheme of dual search algorithm

1. Put $X_1 \in O_n(X^0)$, $i = 1$.
2. In accordance with a rule we choose $X_{i+1} \in O_{n-i}(X^0) \cap O_1(X_i)$. If $X_{i+1} \in S$ then go to step 3; else $i = i + 1$ and repeat the step.
3. $X_{opt} = X_{i+1}$.

From these schemes we can see that a primary algorithm finds a limiting point of the feasible area, but a dual algorithm finds a boundary point which may be not a limiting one. So a primary algorithm finds a better solution then dual in most cases for problems with a connected set of feasible solutions. If we will use a primary algorithm for a problem with a unconnected feasible area then solution received in result may be far from the optimal because feasible and unfeasible solutions will rotate in a path of increasing of the objective function. For these cases a dual algorithm is more useful, because this rotation does not play any role for it. For improving the solution that given by the dual algorithm, it is recommends to apply the corresponding primary algorithm. Such improving is very significant in practice.

Boundary point search algorithms considered below differs by only a rule of choice of a next point in step 2 of the total schemes.

Rule 1. Random search of boundary points (RSB)

A point X_{i+1} is chosen by random way. Each point in the next step can be chosen with equal probabilities. For real-world problems these probabilities can be not equal but they are calculated in the basis of problem specific before search starts.

Rule 2. Greedy algorithm

A point X_{i+1} is chosen accordance with the condition

$$\lambda(X_{i+1}) = \max_j \lambda(X^j),$$

where $X^j \in O_i(X^0) \cap O_1(X_i) \cap S$ for a primary algorithm and $X^j \in O_{n-i}(X^0) \cap O_1(X_i)$ for a dual one.

The function $\lambda(X)$ is chosen from problem specific, for example:

- the objective function $\lambda(X) = C(X)$,
- specific value $\lambda(X) = C(X)/A(X)$ (for only constraint) and so on.

Rule 3. Adaptive random search of boundary points (ARSB)

A point X_{i+1} is chosen by random way in accordance with a probability vector

$$P^i = (p_1^i, p_2^i, \dots, p_J^i),$$

where J is the number of points from which choice is made.

$$p_j^i = \frac{\lambda(X^j)}{\sum_{l=1}^{J} \lambda(X^l)}, \quad j = \overline{1, J},$$

where $X^j \in O_i(X^0) \cap O_1(X_i) \cap S$ for a primary algorithm and $X^j \in O_{n-i}(X^0) \cap O_1(X_i)$ for a dual one.

ARSB is an addition to the greedy algorithm.

Rule 4. Modificated random search of boundary points (MRSB)

A point X_{i+1} is chosen accordance with the condition

$$\lambda(X_{i+1}) = \max_r \lambda(X^r),$$

where X^r are points chosen accordance with the rule 1, $r = \overline{1, R}$; R is a algorithm parameter.

A greedy algorithm is regular algorithm, so it finds equivalent solutions under restart from a certain point. Other algorithms can be started several times and the best solution can be selected from found solutions. Run time of each algorithm start is constrained by

$$T \le \frac{n(n+1)}{2},$$

but average run time of a greedy algorithm and ARSB is significant larger than for other because they looks over all point of the next level in each step in distinction from RSB and MRSB which look over only one and R points correspondingly in each step in the dual scheme.

Further we consider applying the described algorithms for one of real-world problems with large dimension and unconstrained set of feasible solutions.

3. Foundry Branches Production Capacity Optimization

Production of different kind is produces in foundry branches (FB). There is specialization in every FB by kind of production which can be produced by its foundry machines (FM). There is a quantity of orders for production. To each order there corresponds volume, a kind of production and term of performance. The kind of production is characterized by productivity for change (only 3 changes in day). Replacement made on FM production demands its recustomizing borrowing one change. It is necessary to load thus FB capacities that orders were carried out all, production was made in regular more intervals in time, and the number of recustomizings of FM equipment was minimal.

Input data:

I is a number of day for planning;

J is a number of FB;

K_j is a number of FM in *j*-th FB, $j = \overline{1,J}$;

L is a number of orders for production that produces on FM (and corresponding number of production kinds);

V_l is productivity of *l*-th kind of production for change on FM, $l = \overline{1,L}$;

T_l is term of performance of *l*-th order (for production of *l*-th kind) on FM, $l = \overline{1,L}$;

W_l is volume of *l*-th order (for production of *l*-th kind) on FM, $l = \overline{1,L}$;

z_{jl} characterizes specialization of FB:

$$z_{jl} = \begin{cases} 1, \text{ FM of } j\text{-th FB can make production} \\ \qquad\qquad\qquad \text{of } l\text{-th kind,} \\ 0, \text{ otherwise;} \end{cases}$$

α is the factor of rigidity of restriction on demand of uniformity of production on days, $0 < \alpha < 1$.

Variables:

For model construction introduce following binary variables:

$Y = \{y_{ijkl}\} \in B_2^n$,

$X = \{x_{ijkl}\} \in B_2^n$,

where $B_2 = \{0,1\}$, $B_2^n = B_2 \times B_2 \times \cdots \times B_2$ is a set of binary variables.

$$y_{ijkl} = \begin{cases} 1, \text{production of } l\text{-th kind is made} \\ \quad \text{in } i\text{-th day on } k\text{-th FM of } j\text{-th FB,} \\ 0, \text{otherwise;} \end{cases}$$

$$x_{ijkl} = \begin{cases} 1, \text{production of } l\text{-th kind is started to make} \\ \quad \text{in } i\text{-th day on } k\text{-th FM of } j\text{-th FB,} \\ 0, \text{otherwise;} \end{cases}$$

Total dimension of a binary vector *Y* (and *X*) is

$$n = I \cdot L \cdot \sum_{j=1}^{J} K_j .$$

Remarks:

1. For $\forall i, j, k, l$:

$$x_{ijkl} \le y_{ijkl} .$$

2. For $\forall i, j, k, l$:

$$x_{ijkl} = y_{ijkl} \cdot (1 - y_{i-1,jkl}) \quad (5)$$
$$(y_{0jkl} = 0 \ \ \forall j, k, l).$$

3.1. Optimization Model

1. *The objective function and the main constraints*

$$C(X) \to \min , \quad (6)$$

$$A_l^1(Y) \ge W_l , \ l = \overline{1, L} , \quad (7)$$

$$A_i^2(Y) \ge \alpha \cdot W^I , \ i = \overline{1, I} , \ \alpha \in (0,1) , \ W^I = \sum_{l=1}^{L} W_l \Big/ I , \quad (8)$$

where $C(X) = \sum_{i=1}^{I} \sum_{j=1}^{J} \sum_{k=1}^{K_j} \sum_{l=1}^{L} x_{ijkl} = \sum_{i=1}^{I} \sum_{j=1}^{J} \sum_{k=1}^{K_j} \sum_{l=1}^{L} y_{ijkl} \cdot \left(1 - y_{i-1,jkl}\right),$

$$A_l^1(Y) = \sum_{i=1}^{T_l} \sum_{j=1}^{J} \sum_{k=1}^{K_j} V_l \cdot y_{ijkl} \cdot \left(2 + y_{i-1,jkl}\right),$$

$$A_i^2(Y) = \sum_{j=1}^{J} \sum_{k=1}^{K_j} \sum_{l=1}^{L} V_l \cdot y_{ijkl} \cdot \left(2 + y_{i-1,jkl}\right),$$

$$y_{0jkl} = 0 \ \ \forall j, k, l .$$

2. *The additional constraints*

$$\sum_{l=1}^{L} y_{ijkl} \le 1 \ \ (\sum_{l=1}^{L} x_{ijkl} \le 1), \ i = \overline{1, I} , \ j = \overline{1, J} , \ k = \overline{1, K_j} , \quad (9)$$

$$y_{ijkl} \le z_{jl} \ \ (x_{ijkl} \le z_{jl}), \ i = \overline{1, I} ,$$
$$j = \overline{1, J} , \ k = \overline{1, K_j} , \ l = \overline{1, L} . \quad (10)$$

3.2. Model Properties

1. There are two spaces of binary variables (denote their by B^X and B^Y) corresponding vectors *X* and *Y*. For each point $Y \in B^Y$ a unique point $X \in B^X$ corresponds, components of which are determined by relation (5). Several points $Y \in B^Y$ (with different value of constraint function) can correspond to the point $X \in B^X$.

2. The objective function (6) is linear and unimodal monotonic in space B^X with the minimum point $X^0 = (0,...,0)$. In space B^Y the objective function is quadratic and unimodal nonmonotonic with the minimum point $Y^0 = (0,...,0)$.

3. The constraint function (7) and (8) in space B^Y are quadratic and unimodal monotonic pseudo-Boolean functions with the minimum point $Y^0 = (0,...,0)$. In space B^X the constraint functions unequivocally are not certain.

4. The feasible set in spaces B^X and B^Y is limited from above by $I \cdot \sum_{j=1}^{J} K_j$ -th level of the minimum point (X^0 and Y^0) according to the constraint (9). In space B^Y this level corresponds to the case when production is produces on each BM in every day.

5. The feasible set is an unconnected set in general case (in space B^Y).

Thus the problem solution is defined completely by the variables *Y*, but it does not hold for the variables *X*. But the objective function from *X* has good constructive properties so that optimum search on *X* is more efficient then on *Y*. As the constraint function (7), (8) from *X* are not defined, we should find values of these functions from the corresponding point *Y*. There are perhaps several such points

$$X \to Y_1, Y_2, ..., Y_H,$$

in some of them the solution may be feasible but in other not. As the constraint functions are monotonic here then for a certain *X* we should choose such *Y* that belongs to the most possible level (with the most values of the functions):

$$Y = \arg \max_{Y_h, h=\overline{1,H}} \left(\sum_{i,j,k,l} y_{ijkl}^h \right), \text{ where } Y_h = (y_{1111}, ..., y_{IJK_JL}).$$

One of algorithms of this transformation is presented below.

Algorithm 1 of transformation X to Y

1. Put $N_{jk} = 0$, $j = \overline{1,J}$, $k = \overline{1,K_j}$; $i = 1$.
2. For $j = \overline{1,J}$, $k = \overline{1,K_j}$, $l = \overline{1,L}$ do: if $x_{ijkl} = 1$ then $N_{jk} = l$.
3. For $j = \overline{1,J}$, $k = \overline{1,K_j}$, $l = \overline{1,L}$ do: if $N_{jk} = l$ then $y_{ijkl} = 1$ else $y_{ijkl} = 0$.
4. If $i < I$ then $i = i + 1$ and go to step 2.

At the same time the solution *Y* received from the found best vector X_{opt} by this way may corresponds to situation when a quantity of let out production is higher than the requisite value (this does not contradict to the constructed model but this can influent on uniformity of capacities loading which is optimized on the next stage). So when the first stage of search has ended, we should define Y_{opt} by the rule

$$X_{opt} \to Y_1, Y_2, ..., Y_H,$$

$$Y_{opt} = \arg \min_{Y_h : A_l^1(Y_h) \geq W_l, l=\overline{1,L}} \left(\sum_{i,j,k,l} y_{ijkl}^h \right).$$

In this case the transformation algorithm has some differences from the previous.

Algorithm 2 of transformation X to Y

1. Put $N_{jk} = 0$, $j = \overline{1,J}$, $k = \overline{1,K_j}$.

1a. Put $y_{ijkl} = 0$, $i = \overline{1,I}$, $j = \overline{1,J}$, $k = \overline{1,K_j}$, $l = \overline{1,L}$; $i = 1$.

2. For $j = \overline{1,J}$, $k = \overline{1,K_j}$, $l = \overline{1,L}$ do: if $x_{ijkl} = 1$ then $N_{jk} = l$.

3. For $j = \overline{1,J}$, $k = \overline{1,K_j}$, $l = \overline{1,L}$ do: if $N_{jk} = l$ and $A_l^1(Y) < W_l$ then $y_{ijkl} = 1$.

4. If $i < I$ then $i = i + 1$ and go to step 2.

Here the condition $A_l^1(Y) < W_l$ is added in the step 3, and the step 1a is added also for possibility of this calculation. There are no any needs in this transformation during the search. It is necessary only for determining result Y_{opt}.

3.3. Optimization Algorithms

The dual algorithms RSB, greedy and MRSB have been used for solving the problem. The algorithm ARSB has not been considered for this problem because of its excessive large run time by frequent start. One start of ARSB can very rarely give a solution which is better than the solution given by the greedy algorithm. The start point of search is the point of the unconstraint minimum of the objective function $X^0 = (0,...,0)$. A found solution has been improved by the corresponding primary algorithm.

Moreover, the problem has been solved by the genetic algorithm (GA). To realize GA we have chosen a scheme that effective worked for multiple solving other combinatorial optimization problems.

Results of the experiments shows that the most effective algorithms (by precision and run time) from the considered ones for this problem are the greedy algorithm and MRSB. The other algorithms under hard constraints on the variables do not find any accessible solution at all. It is a sequel of problem specific: a large amount of different constraints and, as a result, a comparative small accessible region.

The average results of solving 10 problems of month planning capacity loading are presented in the table 1. The average values of input data:

$I = 31$, $J = 3$, $K_1 = 12$, $K_2 = 9$, $K_3 = 7$, $L = 36$, $\alpha = 0.5$,

$V_l \in [40,50]$, $W_l \in [20,25000]$.

Herewith the total dimension of the binary vector is $n = 31248$.

The number of algorithm starts L has been chosen so that the run time nearly equals to the run time of one start of the greedy algorithm. In this case the run time equals to $T \approx 8 \cdot 10^5$.

Table 1. Optimization results

Algorithm	Number of starts *L*	Found solution C_{opt}
RSB	2000	Not found
Greedy	1	49
MRSB, *R*=1000	12	47
MRSB, *R*=100	60	52
MRSB, *R*=10	200	Not found
GA*	-	Not found

* - GA parameters: turnir selection with the turnir value 5, population value 100, the largest number of generations 8000, mutation probability 0.0001.

MRSB is a more flexible procedure in compare with the greedy algorithm as the first one allows selecting the parameters *L* and *R* that influence on algorithm run time and solution precision. The greedy algorithm does not allow that possibility and run time may be overmuch large under high dimensions. What about their efficiency, precision of the found solutions differs unessentially under nearly equivalent run time.

So, the algorithms of boundary point search show high efficiency for solving the pseudo-Boolean optimization problem with unconnected accessible region. The most efficient algorithms for the considered problem are the dual algorithms MRSB and greedy.

Acknowledgments
The results were obtained within the project No. 346 of the state task No. 2014/211 of the Ministry of Education and Science of the Russian Federation.

References

[1] Aarts E H L and Lenstra J K 1997 Local Search in Combinatorial Optimization *Wiley-Interscience Series in Discrete Mathematics and Optimization* (John Wiley & Sons Ltd)

[2] Papadimitriou C H and Steiglitz K 1998 *Combinatorial Optimization* (New York: Dover Publications)

[3] Crama Y and Hammer P L 2011 *Boolean Functions: Theory, Algorithms, and Applications* (New York: Cambridge University Press) p 687

[4] Masich I S 2013 *Search Algorithms for Conditional Pseudoboolean Optimization* (Krasnoyarsk: Siberian State Aerospace University) p 160

[5] Björklund H, Sandberg S and Vorobjov S 2002 Optimization on Completely Unimodal Hypercubes *Technical report* 2002-018 p 39

[6] Wegener I and Witt C 2002 On the Optimization of Monotone Polynomials by Simple Randomized Search Heuristics *Technical Report* 1433-3325

[7] Hammer P L and Rudeanu S 1968 *Boolean Methods in Operations Research and Related Areas* (Springer) p 310

[8] Boros E and Hammer P L 2002 Pseudo-Boolean Optimization *Discrete Applied Mathematics* **123(1-3)** pp 155-225

[9] Allemand K, Fukuda K, Liebling T M and Steiner E 2001 A polynomial case of unconstrained zero-one quadratic optimization *Math. Program.* A **91** pp 49-52

[10] Boros E and Hammer P L 2008 A Max-Flow Approach to Improved Lower Bounds for Quadratic Unconstrained Binary Optimization (QUBO) *Discrete Optimization* **5(2)** pp 501-529

[11] Foldes S and Hammer P L 2000 Monotone, Horn and Quadratic Pseudo-Boolean Functions *Journal of Universal Computer Science* **6(1)** pp 97-104

[12] Goldberg D E 1989 *Genetic algorithms in search, optimization, and machine learning* (Addison-Wesley)

[13] Schwefel H P 1995 *Evolution and Optimum Seeking* (N.Y.: Whiley Publ.) p 612

[14] Antamoshkin A N 1989 *Regular optimization of pseudo-Boolean functions* (Krasnoyarsk: Krasnoyarsk university publ.) p 284

[15] Antamoshkin A N and Masich I S 2007 Pseudo-Boolean optimization in case of unconnected feasible sets *Models and Algorithms for Global Optimization* vol 4 of the series *Optimization and Its Applications* ed A Torn and J Zilinskas (Springer) pp 111-122

[16] Antamoshkin A N and Masich I S 2006 Heuristic search algorithms for monotone pseudo-boolean function conditional optimization *Engineering&automation problems* vol 5, N 1 pp 55-61

IOP Conf. Series: Materials Science and Engineering **122** (2016) 012002 doi:10.1088/1757-899X/122/1/012002

Selection of logical patterns for constructing a decision rule of recognition

A N Antamoshkin[1,2] and I S Masich[1,3]
[1] Department of System Analysis and Operation Research, Siberian State Aerospace University named after academician M. F. Reshetnev, Krasnoyarsk, Russia
[2] Department of Economics and Information Technology of Management, Siberian Federal University, Krasnoyarsk, Russia
[3] Department of Informatics and Mathematical Modeling, Krasnoyarsk State Agrarian University, Krasnoyarsk, Russia

E-mail: i-masich@yandex.ru

Abstract. We investigate an aspect of the construction of logical recognition algorithms – selection of patterns in the set of found patterns in the data.
We consider the recognition problem for objects described by binary attributes and divided into two classes. A result of performance the procedure of searching patterns on the training set (a set of input data) is a number of patterns found. The question is to select some patterns from their total number to form a decision rule. That can not only reduce size of the decision rule, but also to improve recognition.
One way to make a selection of patterns is select a subset of patterns that are needed to cover all objects of the training sample. This problem is formulated as an optimization problem. The resulting optimization model represents a problem of conditional pseudo-Boolean optimization, in which the objective function and the constraints functions are unimodal monotone pseudo-Boolean functions.
Another way is to make the selection of such patterns, which when used together will increase separating capacity of the decision rule. As a criterion for the formation of the decision rule is considered the width of the separation margin. One more way is the selection supporting objects, rules are formed on the basis of which.
Selection of logical patterns, which is made in accordance with the proposed approach, can significantly reduce the number of patterns and simplify the decision rule, almost without compromising the accuracy of recognition. This makes the decision rule clearer, and the results more interpretable. It is necessary to support decision making for recognition.

Statement of the pattern selection problem

Currently pretty effective classification algorithms are developed for the solution of problems of diagnostics and prediction, and ably tuning these algorithms solve the problems with high accuracy. But in the practical application of such algorithms, we often are interested in the question of evidence and interpretability of results. For decision making a model in explicit form is required, such a model in which the calculated decisions are justified and based on available data. In this paper we construct a model for decision making which consists of a set of logical rules that describe patterns in a studied phenomenon or a system. The main task is to identify these patterns and lead to a form in which they will be used to construct a decision making model. Identification of pattern on the basis the available set of data is a complex computational task that requires efficient algorithmic support and software implementation. Even so constructed classifiers can effectively solve practical problems [1, 2], including those in the aerospace industry [3, 4].

The process of formation of decision rules is accompanied by solving problems of selection of the best alternatives in accordance with some criterion. Formalization of this process in a number of problems in combinatorial optimization generates flexible and efficient algorithm for logical analysis for data classification.

Let us consider the problem of recognition of objects described by binary attributes and divided into two classes $K = K^+ \cup K^- \subset \{0,1\}^n$. An object (or observation) $X \in K$ is describe by binary vector $X = (x_1, x_2, ..., x_n)$ and can be can be represented as a point in the hypercube of binary attribute space B_2^n.

A term which covers at least one observation of a certain class and does not cover any observation of another class is called a pattern or a rule. That is, the pattern corresponds to the cube having non-empty intersection with one of the sets (K^+ or K^-) and an empty intersection with another set (K^- or K^+ respectively). A pattern *P*, which is disjoint with K^-, will be called positive, and a pattern *P'*, which is disjoint with K^+, will be called negative. Patterns are the elementary building blocks for rule-based algorithms.

Assume that as a result of the patterns search procedures on the training set we found some positive patterns *Pi*, *i*=1,…,*p*, and some negative patterns *Nj*, *j*=1,…,*n*.

The discriminant (decision) function can be given by the expression

$$D(a) = \frac{1}{p}\sum_{i=1}^{p} P_i(a) - \frac{1}{n}\sum_{j=1}^{n} N_j(a)$$

for some observation *a*,

where *Pi*(*a*)=1 if the pattern *Pi* cover the observation *a* and *Pi*(*a*)=0 in another case. The same goes for *Nj*(*a*).

The algorithms for finding patterns are described in [5]. In particular, that is the algorithms that search patterns by basing on some observation of the training sample. Therefore, as a result of such search a large number of patterns can be recorded, up to the number of training observations. Some of these patterns, however, may coincide. In solving many problems there is a question of selection patterns from the set of all patterns to form a decision rule. That can not only reduce the size of classifier, in some cases that can improve recognition.

The main advantage that the rule-based algorithms provide for solving real-world problems is clarity of the process of recognition on the resulting model for new observations. All identified patterns are presented explicitly. However if the number of these patterns is large (more than 10-15) then the recognition algorithm becomes difficult to interpret. In this regard, let us explore some methods of selection from the total number of patterns discovered.

Minimizing the number of patterns

Let us introduce the variables that determine whether the pattern presents in the decision function.

$$x_i = \begin{cases} 1, P_i \text{ presents in the decision function,} \\ 0, \text{ in another case.} \end{cases}$$

$$y_j = \begin{cases} 1, N_j \text{ presents in the decision function,} \\ 0, \text{ in another case.} \end{cases}$$

One way to make a selection of patterns is select a subset of patterns, which are needed to cover all the training observations [6]. Thus each object of the training sample must be covered by at least one pattern. Using the variables introduced above, this condition can be written as

$$\sum_{i=1}^{p} x_i P_i(a) \geq 1 \text{ for any } a \in K^+,$$

$$\sum_{j=1}^{n} y_j N_j(a) \geq 1 \text{ for any } a \in K^-.$$

To improve the robustness of the algorithm the number 1 in the right side of inequalities should be replaced by a positive integer *d*. In this case, each object of the training sample must be covered by *d* patterns.

Thus, we have the following problem of minimizing the number of patterns used in the decision rule

$$\sum_{i=1}^{p} x_i + \sum_{j=1}^{q} y_j \to \min$$

subject to

$$\sum_{i=1}^{p} x_i P_i(a) \geq d \text{ for any } a \in K^+,$$

$$\sum_{j=1}^{n} y_j N_j(a) \geq d \text{ for any } a \in K^-.$$

The obtained optimization model represents a problem of conditional pseudo-Boolean optimization, in which the objective function and the constraint functions are unimodal and monotone pseudo-Boolean functions [7]. To solve the problem we use the approximation algorithms for the conditional pseudo-Boolean optimization, based on search for the optimal solution among the boundary points of the feasible region [8].

In order to evaluate the effect of reducing the number of patterns in the final rule on recognition accuracy, a series of experiments was held on recognition and prediction problems. Patterns search was made on the basis of the optimization model [9], which allows to find the maximum patterns, that is patterns with the highest coverage of observation of a certain class. Each data sample has been divided into two parts: a learning sample and a test sample. On the basis of each observation of the training sample a pattern was found. A comparison of quality recognition decision rules is made. One decision rule is built from the complete set of patterns and another rule from a reduced set obtained by solving the optimization problem described above.

In the experiments we used the following recognition problems [10]:

breast-cancer – the problem of breast cancer diagnosis, sample size: 699 observation described by 9 multi-type attributes (80 binary attributes obtained by binarization);

wdbc – the problem of breast cancer diagnosis, sample size: 569 observation described by 30 multi-type attributes (120 binary attributes);

hepatitis – the problem of hepatitis diagnosis, sample size:155 observation described by 19 multi-type attributes (37 binary attributes);

spect – data on cardiac computed tomography, sample size: 80 observation described by 22 binary attributes.

The results are shown in Table 1.

Table 1. Recognition results

Recognition problem	Set of patterns	Number of positive patterns	Number of negative patterns	Recognition accuracy for positive observations	Recognition accuracy for negative observations
breast-cancer	complete set	419	209	0.97	0.91
	reduced set	12	14	0.97	0.88
wdbc	complete set	291	163	0.94	0.98
	reduced set	9	11	0.92	0.96
hepatitus	complete set	27	97	0.8	0.85
	reduced set	7	7	0.8	0.81
spect	complete set	38	34	1	0.83
	reduced set	7	8	1	0.83

As can be seen from the results, the application of a decision rule based on a reduced set of patterns in some problems leads to a slight decrease in recognition quality. But at the same time this accompanied by a significant decrease in the number of patterns to be used for decision making, and that positively affects to clarity of the decision obtained.

Maximizing the separation margin

Another method is to make a selection of such patterns, which when used together will increase the separating ability of the decision rule.

As a criterion for the formation of the decision rule, consider the width of the "separating margin".

$$\min\{D(a): a \in K^+\} - \max\{D(a): a \in K^-\},$$

where $D(a) = \frac{1}{p}\sum_{i=1}^{p} P_i(a) - \frac{1}{n}\sum_{j=1}^{n} N_j(a)$ for some observation a.

We take into account the presence of outliers that may be present in real-world problems. For this purpose we introduce a variable

$$z^a = \begin{cases} 1, a \text{ is taken as an outlier,} \\ 0, \text{in another case.} \end{cases}$$

Then the problem of pattern selection can be written as follows.

$$v^+ + v^- - C\sum_{a\in K} z^a \cdot |b^a| \to \max,$$

$$\text{where } v^+ = \min\{D'(a): a \in K^+, z^a = 0\},$$

$$v^- = \min\{-D'(a): a \in K^-, z^a = 0\},$$

$$D'(a) = \frac{\sum_{i=1}^{p} x_i P_i(a)}{\sum_{i=1}^{p} x_i} - \frac{\sum_{j=1}^{n} y_j N_j(a)}{\sum_{j=1}^{n} y_j},$$

$$b^a = \begin{cases} v^+ - D'(a), a \in K^+, \\ v^- + D'(a), a \in K^-. \end{cases}$$

Algorithms for solving such optimization problems are given in [11].

Decomposition of training sample in identifying patterns

Considered in [12, 13] methods to search for patterns suggest use as a "support" point an observation of the training sample (precedent), a partial overlapping of the properties of which can be found in other observations of the same class. The method described above requires to use a large number of support observations (possibly all observations of the training sample) to obtain patterns, and then to carry out the selection from the found patterns.

Consider another method comprising selection of these support objects. The entire set of observations of training sample for a certain class, for example K^+, can be divided into groups of observations so that the observations were quite similar in each group:

$$K^+ = K_1^+ \cup K_2^+ \cup \ldots \cup K_k^+$$

For this purpose, it can be used the k-means algorithm, as a result of which we obtain a set of centroids $c_1, c_2, \ldots, c_k$, so that the following rule holds:

$$a \in K_j^+, \text{ if } \|a - c_j\| < \|a - c_i\|$$
$$\text{For all } i = 1,2,\ldots,k,\ i \neq j,$$

where K_j^+ is the set of observations included in the cluster with the centroid c_j.

These centroids can be used as support observations to identify logical patterns.

Described approach can significantly reduce the complexity of the logic recognition algorithm, producing a selection of observations used as support observations when searching for patterns.

Let us consider the results of using this approach in relation to the problem of predicting myocardial infarction complications: atrial fibrillation (AF) and ventricular fibrillation (VF) [14]. To find centroids the k-means algorithm was used in software applications Weka [15]. To search for patterns and estimate the accuracy of the constructed decision rule the software of authors was used.

The sample for the AF problems consisted of 184 positive and 184 negative observation described by 112 multi-type attributes. The number of binarized attributes was 215. For each class 15 centroids was allocated, which were used to search for patterns.

The sample for the VF problems consisted of 80 positive and 80 negative observation described by 112 multi-type attributes. The number of binarized attributes was 200. For each class 10 centroids was allocated, which were used to search for patterns.

10% of observations of the sample were allocated for testing the resulting decision rule. Recognition results are shown in Table 2.

Table 2. Comparison of results of recognition

Recognition problem	Set or patterns	Number of positive patterns	Number of negative patterns	Recognition accuracy for positive observations	Recognition accuracy for negative observations
AF	complete set	165	165	0.7	0.79
	reduced set	15	15	0.68	0.77

VF	complete set	72	72	0.87	0.71
	reduced set	10	10	0.9	0.88

As a result of the decomposition of observations of training sample and the corresponding selection of observations used as support observations for finding patterns, we obtain simplification of the decision rule - the number of patterns used in the final rule decreases in 7-10 times. At the same time for some problems we even observe increasing in accuracy of recognition of test observations.

Conclusion

To summarize, it must be concluded that the selection of logical patterns, produced in accordance with certain criteria, can significantly reduce the number of pattern used in the classifier and simplify the decision rule, only slightly reducing the accuracy of recognition. In solving a number of real-world problems of recognition and prediction, great importance is the interpretability of the obtained solutions and the ability to justify them on the basis of rules and patterns, which, in its turn, are based on precedents in the form of observations in the data sample. Therefore, the use approaches described in this paper is useful for solving such problems.

Acknowledgments

The results were obtained within the project No. 346 of the state task No. 2014/211 of the Ministry of Education and Science of the Russian Federation.

References

[1] Alexe G, Alexe S, Hammer P L, Liotta L, Petricoin E and Reiss M 2004 Ovarian Cancer Detection by Logical Analysis of Proteomic Data *Proteomics* **4(3)** 766–783

[2] Axelrod D, Bonates T, Hammer P L and Lozina I 2004 From Diagnosis to Therapy via LAD *Invited Lecture at INFORMS Annual Meeting* (Denver: CO)

[3] Dupuis C, Gamache M and Páge J F 2012 Logical analysis of data for estimating passenger show rates in the airline industry *Journal of Air Transport Management* **18** 78–81

[4] Esmaeili S 2012 Development of equipment failure prognostic model based on logical analysis of data *Master of Applied Science Thesis, Dalhousie University, Halifax, Nova Scotia, July 2012*

[5] Masich I S 2009 Combinatorial Optimization in Classification Problem *Control Systems and Information Technology* **1.2(35)** 283-288

[6] Hammer P L and Bonates T O 2006 Logical analysis of data - An overview: From combinatorial optimization to medical applications *Annals of Operations Research* **148**

[7] Antamoshkin A N and Masich I S 2007 Pseudo-Boolean optimization in case of unconnected feasible sets *Models and Algorithms for Global Optimization* vol 4 of the series *Optimization and Its Applications* ed A Torn and J Zilinskas (Springer) pp 111-122

[8] Antamoshkin A N and Masich I S 2003 Efficient algorithms for constrained optimization of monotone pseudo-Boolean functions *Vestnik SibSAU* **4** 60-67

[9] Golovenkin S E, Gulakova T K, Kuzmich R I, Masich I S and Shulman V A 2010 Model of logical analysis for solving the problem of predicting myocardial infarction *Vestnik SibSAU* **4(30)** 68-73

[10] Lichman M 2013 UCI Machine Learning Repository [http://archive.ics.uci.edu/ml] *Irvine, CA: University of California, School of Information and Computer Science*

[11] Masich I S 2006 The heuristic algorithms of boundary points search for an constraint pseudo-boolean optimization problem *Vestnik SibSAU* **1(8)** 39-43

[12] Alexe G and Hammer P L 2006 Spanned patterns for the logical analysis of data *Discrete Applied Mathematics* **154** 1039–1049

[13] Guoa C and Ryoo H S 2012 Compact MILP models for optimal and Pareto-optimal LAD patterns *Discrete Applied Mathematics* **160** 2339–2348

[14] Golovenkin S E, Gorban A N and Shulman V A 1997 Complications of myocardial infarction: a database for testing recognition and prediction *Preprint* 1997.6 (Krasnoyarsk: Inst. of Computational Modeling SD RAS)

[15] Waikato Environment for Knowledge Analysis. Weka 3: Data Mining Software in Java http://www.cs.waikato.ac.nz/ml/weka

Multi-agent automation system for monitoring, forecasting and managing emergency situations*

O A Antamoshkin[1,2], O A Antamoshkina[2,3], N A Smirnov[1]

[1] Siberian State Aerospace University named after academician M. F. Reshetnev, Krasnoyarsk, Russia
[2] Siberian Federal University, Krasnoyarsk, Russia
[3] V.N. Sukachev Institute of Forest SB RAS, Krasnoyarsk, Russia

E-mail:oleslav24@gmail.com

Abstract. The paper outlines the general concept of multi-agent approach to develop the automation system for monitoring, forecasting and managing emergency situations and its models and algorithms included.

Introduction. Emergency situations response and prevention efficiency primarily depends on the availability of required information on controlled objects and processes and its quick transition into instructions, plans, projects and actions. Such terms as information systems, automation management system, data base management system, etc. are appeared to be determined when considering relations between management and information.

The main objectives of monitoring and forecasting are observation, control and prediction of hazardous processes and nature phenomena, technosphere and external disturbing factors (armed conflicts, terrorist attacks, etc.) acting as a source of emergencies. They are also committed to track development dynamics of emergency situations and define their scale within the framework of measures taken in prevention and elimination of hazards.

Known solutions. The information is the key resource that directly enhances efficient activities [1, 2]. The management theory reasonably describes information as a unity of data on all the changes occurring in the system and its environment that decreases our knowledge uncertainties pertaining to specific object. It enables to organize data exchange between people, human and machine, machine and machine. Information is used as a resource to execute work functions as well as a mean of business communications arranged in the process of various data transmission.

It is obvious that such system is impossible to be developed on the single platform because of its fragmentation, constant multi-task increase and inter-agency incoherence. Such system is required to be considered as a complex geographically-distributed hierarchical structure processing information, material, power and other possible resource flows to make decisions in terms of collection and semantic analysis of information received from multiple heterogenic channels. Nowadays possessors of its certain components are successfully applying functional software with no intentions to reject implemented solutions. As practice of establishing the operating control system within even EMERCOM of Russia shows, such projects may be prolonged for a certain period of time without any results expected.

The issue was frequently described in works of A.A. Vavilov, V.I. Varshavskiy, V.A. Vittikh, V.M. Glushkov, T.A. Gavrilova, V.I. Gorodetskiy, L.A. Kalinichenko, A.V. Kostrov,

* The results were obtained within the project No. 346 of the state task No. 2014/211 of the Ministry of Education and Science of the Russian Federation.

O.I. Larichev, A.A. Letichevskiy, V.V. Lipaev, V.E. Marley, J. Martin, M. Minskiy, G.S. Osipov, E.B. Pesikov, G.S. Pospelov, D.A. Pospelov, E.B. Popov, K.A. Pupkov, B.L. Sovetov, V.L. Stefatok, E.A. Trakhtengerts, B.F. Fomin, V.F. Khoroshevskiy, I.G. Chernorutskiy, R. Shenk, E.L. Yushenko, S.A. Yakovlev, A.I. Yashin.

The research work is based on multi-agent approach application enabling to unite current parts of information environment into unified system of data exchange step by step in order to develop distributed and integrated decision support for Russian System of Prevention and Response to ES.

The crucial objective of multi-agent management system is to distribute the agents. The task is performed by optimized program agents distribution in different sites of computer networks. In this case, optimality is characterized by shrinking time for data execution needed for the decision support system. This objective is appeared to be complicated due to a great number of limiting factors available while distributing agents. Based on the concept of developing distributed computer networks for decision making in extreme situations, it is noted that technical resources are owned by various organizations and institutions thus blocking further system establishing on the unified techno and program platform.

Proposed solutions. Nevertheless, by studying different information systems included into distributed computer networks for decision making in extreme situations, the whole range of typical solutions somehow applied to develop each of them can be distinguished. In spite of diversity in approaches and their development, the network infrastructure in most cases counts on solutions suggested by CISCO Company that is considered to be acknowledged leader in telecommunication facility production (it is estimated that the facilities of this company provide of about 80 % of web-traffic). It is explained by high standards for data transmission security and certification required for cryptographic solutions –program and hardware-software systems developed by «Infotecs» Company. The program part of its network equipment is based on Cisco IOS or adapted version of Linux operating system (products of «Infotecs» Company). Nowadays, both determined operating systems support OS capability above Linux core and provide high availability and integration of services.

To achieve acceptable level of standardization and unification of NetWare Management Agents, it is suggested to place them exactly on the base of communication equipment of these vendors.

The additional advantage of such placement is low work load of telecommunication server computing resources that is confirmed by statistical analysis of 170 units of Siberian segment of distributed computer network for decision making in extreme situations. The analysis results showed that daily average processor utilization comprised not more than 12,3% out of the highest possible level.

The range of management agents observed enables to track raw data collection from the source units, monitor workload of distributed computer network resources, coordinate computing tasks assignment between them, manage data stores and enhance efficiency of data communication channel usage. This range of agents is managed to be influenced by signals of out-of-order requests on data and values production from the decision support system.

It should be noted that the goal we are going to achieve is similar to Zeno's paradox due to constant development of distributed computer network elements. However, the system solution approach allows downsizing the distance to «tortoise» rather quickly. Along with management agents distribution, the template solutions for any state-of-the-art information systems development are recommended to be considered. The majority of state-of-the-art information systems is aimed at collecting and processing mass data. There is also no exception pertaining to the distributed computer network for decision making in extreme situations. The model data base management systems (DBMS) are applied to process and store the information. After statistical data processing, it is possible to make conclusion that the following DBMS tend to prevail in distributed computer network for decision making in extreme situations – MS SQL, Oracle RDBMS with MySQL dominating in a smaller extent to solve various local tasks. Implementation of data exchange agents in the DBMS enables to computerize communication flow in the distributed computer network for more than 70%. Furthermore, retrieving information from the databases will allow minimizing development of raw data collecting agents at the initial stages.

The main requirements applicable to agents-integrators are as follows: knowledge of basic SQL requests (get access to database, add information to the database, change information in database, remove information from the database, etc.), language skills for data message exchange in multi-agent environment, database request understanding, database encoding skills, reporting to managing agents.

In most cases, the analysis of technical and organizational systems is managed to describe such components as mission, vision, strategy and processes. Multi-agent approach application is reported to determine the systems as dynamic ones based on the values.

The dynamic model of discrete process is characterized by means of the following mathematical tool: Petri net, multiqueue system and system dynamics model. Later on, the intelligent agents supplement the models. Consequently, the integration of simulation modelling, expert systems, case and multi-agent modelling is considered to be resolved.

The model of the intelligent agent shall be represented as follows:

Agent=<Name, Purpose, Priority, KB, M_in, M_out, B_scen, Control_O, Chief_A, Sub_A>,

where Name – the name of the agent, Purpose – goals of the agent, Priority – priorities of the agent, KB – database of the agent, M_in – number of incoming messages, M_out – number of outgoing messages, B_scen – performance plan, Control_O – multitude of controlled objects, Chief_A – multitude of dominated agents, Sub_A – multitude of subordinated agents.

The agent is capable to accomplish the following actions:

- Analyze the current situation though external parameters evaluation;
- Detect the situation;
- Manage the knowledge base;
- In case of detecting certain situation, the agent is managed to seek the action script in the knowledge base or set it up independently;
- Make decisions;
- Define and adjust its goals and goals of subordinated agents;
- Control goals achievement;
- Exchange messages.

In order to build the system core, the tool of production systems is applied. The structure of production system is described as follows:

PS=<CSs, KBs, Ms>,

where $CSs=\{RES(t)\}\cup\{MECH(t)\}\cup\{U(t)\}\cup\{G(t)\}$ – current status of resources, facilities, executive instructions, goals (рабочей памяти); KBs – the rule set of agent's activity and transformations (knowledge base); Ms – inference engine consisting of task scheduler and inference machine of the agents' knowledge data base.

The method comprises the following steps:

Step 1. Determination of current time SystemTime=min(Tj), j∈rule.
Step 2. Processing of agent's actions.
Step 3. Algorithms queue scheduling.
Step 4. Execution of rules and goals modification.

To detect the current state of the system and generate control commands, the system refers to consulting program.

The hierarchical structure of the multi-agent system is specified by means of system flow charts with large scale integration level and the object – based structured approach for developing models of the technical and organizational systems suggested by T.A. Gavrilova [3]. The approach based on frame-concepts and conceptual flow charts proposed by A.N. Shvetsov is used as a tool for knowledge formalization [4]. The task description of resources allocation is declared by the author in the research work [5].

As defined previously, the model of intelligent agent in general terms is represented by Agent=<Name, Purpose, Priority, KB, M_in, M_out, B_scen, Control_O, Chief_A, Sub_A>. Furthermore, its model of behavior scenario shall be outlined - B_scen:

B_scen = (MIS, MG, MSR, MA),

Where MIS –model of information space, MG – model of goal - setting, MSR – solution search model, MA – model of active actions. The intelligent agent is managed to make decisions on realization of certain scenario at a time.

From the standpoint of actions execution, the models are considered to be divided into three classes: predetermined finite set of primitive actions, plans multitude, selectable messages and actions in the logical language.

The intelligent agent is suggested to obtain a great number of statical goals $SP=\{sp_j, j=1,\dots,n\}$. There is a certain range of $Rules_j$, $j=1,\dots,n$, that are followed and lead to sp_j. Each agent adheres to certain plan with messages and actions already established within.

The solution search model is designated by the function $SS: SP\rightarrow VO$, where VO – the set of rules of intelligent agent. The model of active actions is defined by the visualization $AM: AS\rightarrow VO$, which selects the rules required to be executed at the moment. Active goals alteration (*AS* multitude) is expected to result in termination or activation of certain rules.

SA(0) shall be estimated as the entry-level configuration of intelligent agent's properties with *lvo* the list of its rules, *SMA(0)* – entry-level configuration of *MA*, *ls* – the list of agent's strategies, *lsg* – the list of statical goals, *ltg* – the list goals prescribed by dominating agent, *lbg* – list of goals transmitted to subordinating agents, *ldag* – the list of dynamic goals, *lag* – the list of active goals. Calculation axiomatic statement for K_{B_scen} at the time t_0 shall be as follows $AA=(SA(0),lvo,SMA(0),ls(0),lsg(0),\emptyset,\emptyset,\emptyset,\emptyset)$.

To simplify transcription of information space V and its status SV, we shall identify:

$lia=\{Name_j,\{A_{Agentj}^{k}, k=1,\dots,m\},j=1,\dots,l\}$,

$lio=\{Name_j,\{A_{Agentj}^{k}, k=1,\dots,m\},j=1,\dots,l\}$,

The independent operation of the agents requires assignment of decision algorithms. The algorithms are exemplified below.

Decision-making on data collection for calculation (Algorithm 1).

1. Define the list of necessary data
2. Estimate the time for all the data collection
3. Arrange data by the receipt time
4. Arrange data by receipt requirements for accurate calculation. Determine minimal (threshold) values for calculation.

5. Delimit the objective on two stages in accordance with Step 4 – approximate calculation and full calculation.
6. Set priority 1 for the data assigned to preliminary calculation and priority 2 for the other one.
7. Proceed with data collection (in parallel with all the sources) in terms of priorities.
8. Send the calculated model to decision-maker upon completion of preliminary calculation.
9. Evaluate time needed to accomplish final calculation. In case it is less than the time left to make decision, continue to collect data.
10. Elaborate final calculation and send it to decision-maker.

Decision on route selection (Algorithm 2).

In this particular case, the route is characterized as not only data transmission but also as data transformation (computing). The instrument of multi-version control suggested in works [6-8] is applicable in this algorithm.

After requesting data transmission to decision-maker:

1. Examine the network and make a list of possible routes.
2. Forecast time and reliability (transfer expectations) of the routes. Subordinated agents perform these calculations.
3. Determine the necessity to launch concurrent processes by utilizing the instrument of multi-version control.
4. Set up data transfer through the routes identified.

Algorithm of data transmission (transmission only) (Algorithm 3).

1. Receive the close-in point of data production from superior agent (the server nearby enabling to transit current data to the data of higher rank).
2. Define the volume of data transmitted to the closest point of data production.
3. Make a list of possible routes.
4. Select the best route by the function value $F(x)=TP*R(t+1)$, where TP- certainly value of adequate amount of data transmission to the end point in accordance with t, $R(t+1)$ – certainly value of transmission channel efficiency at the moment $t+1$ (operational availability).
5. Set up data transmission.

The example of algorithm for distributed geoinformation system (based on Algorithm 1).

1. Define the goal of modeling from the request of decision-maker.
2. Compose a list of data for modeling.
3. Make a list of data resources.
4. Examine data supply sources in terms of readiness for the actual data transmission.
5. Organize schedule for data delivery.
6. If probability estimate of total time needed for data acquisition (T_1), data delivery (T_2), transformation (computing) (T_3) is more than the time assigned for making decision (T_r), it is considered to divide the task on 2 steps: sample calculation and final calculation.
7. Take up the time available for sample calculation (T_{r1}) from inequality $T_{r1} <= T_r$.
8. Outline time for primary data receipt as $T_{1'} <= T_{r'} - T_2 - T_3$.

9. Set out the list of data possible to be delivered during the certain time period in accordance with T_{Γ}.
10. Launch data collection.
11. Inquire managing agents about current and prognostic state of the Network.
12. Settle possible routes.
13. Identify data pools based on the object pools of the model.
14. Launch data transmission for calculation upon closing (collecting all the necessary information for calculation) data pools.

Conclusions. The research activities have resulted in developing methodology for inter-agency intelligent decision support in the Russian System of Prevention and Response to ES which allows upgrading the intelligent level of ES monitoring and forecasting system acting as a functional informative-analytical subsystem of the RSPR. Consequently, the time needed for control solution shall be downsized greatly in conditions of changing information environment. The complex of methods elaborated enables to computerize the development process of corporate, intelligent decision support systems, consider specific features of subject field and type of management tasks solved, minimize expenses on exploitation and enhance efficiency of information resources applied.

References

[1] Nordstrom Kjel, Ridderstrale Jonas, Funky Business. Stockholm School of Economics in St. Petersburg. 2003. p. 165.
[2] Pavlov S. GIS Application for Monitoring and Management of ES. ArcReview. 3(26), 2003.
[3] Chastikov A.P., Gavrilova T.A.., Belov D.L. Development of Expert Systems. CLIPS Environment. Saint-Petersburg: BHV-Petersburg, 2003. 608 p.
[4] Shvetsov A.N. Models and Methods for Developing Corporate, Intelligent Decision Support Systems: PhD thesis: 05.13.01 Saint-Petersburg, 2004. 461 p.
[5] Antamoshkin O.A. Technology of Distributed Computer Network Resources Utilization. Management Systems and Information Technologies. Voronezh: Scientific Book, 2012. Issue 2.2 (48). pp. 220-224.
[6] Kovalev I.V., Tsarev R.Yu., Prokopenko A.V., Soloviev E.V. Considering Realization of Ant Algorithm by Selecting Composition for Multi-version Software of Information Management Systems. Tools and Systems. Management, Control, Diagnostics. 2012. № 2. pp. 1-4.
[7] Kovalev I.V., Slobodin M.Yu., Tsarev R.Yu. Determination of Subject Preferences of Decision-maker in terms of Multi-version Projection of Automatic Management Systems. Management Systems and Information Technologies. 2005. № 1(18). pp. 44–49.
[8] Antamoshkin O.A., Puzanova G.A., Ontugev V.V. Design features of the automated system of expert assessment of information security organizations. Vestnik the Siberian State Aerospace University named by Academician M.F. Reshetnev 3 (49) pp. 4-13.

Landsat imagery evidences great recent land cover changes induced by wild fires in central siberia*

O A Antamoshkina[1,2], N V Trofimova[3], O A Antamoshkin[4,5]
[1] Researcher, V.N. Sukachev Institute of Forest SB RAS, Krasnoyarsk, Russia
[2] Senior researcher, Siberian Federal University, Krasnoyarsk, Russia
[3] Head of the Altai-Sayan Ecoregional Office WWF, Krasnoyarsk, Russia
[4] Candidate of tech. sciences, Associate professor, Siberian State Aerospace University named after academician M. F. Reshetnev, Krasnoyarsk, Russia
[5] Candidate of tech. sciences, Associate professor, Siberian Federal University, Krasnoyarsk, Russia

E-mail: sloa@mail.ru

Abstract. The article discusses the methods of satellite image classification to determine general types of forest ecosystems, as well as the long-term monitoring of ecosystems changes using satellite imagery of medium spatial resolution and the daily data of space monitoring of active fires. The area of interest of this work is 100 km footprint of the Zotino Tall Tower Observatory (ZOTTO), located near the Zotino settlement, Krasnoyarsk region. The study area is located in the middle taiga subzone of Western Siberia, are presented by the left and right banks of the Yenisei river.
For Landsat satellite imagery supervised classification by the maximum likelihood method was made using ground-based studies over the last fifteen years. The results are the identification of the 10 aggregated classes of land surface and composition of the study area thematic map.
Operational satellite monitoring and analysis of spatial information about ecosystem in the 100-kilometer footprint of the ZOTTO tall tower allows to monitor the dynamics of forest disturbance by fire and logging over a long time period and to estimate changes in forest ecosystems of the study area. Data on the number and area of fires detected in the study region for the 2000-2014 received in the work. Calculations show that active fires have burned more than a quarter of the footprint area over the study period. Fires have a significant impact on the redistribution of classes of land surface. Area of all types of vegetation ecosystems declined dramatically under the influence of fires, whereas industrial logging does not impact seriously on it.
The results obtained in our work indicate the highest occurrence of fires for lichen forest types within study region, probably due to their high natural fire danger, which is consistent with other studies. The least damage the fire caused to the wetland ecosystem due to high content of moisture and the presence of a large number of fire breaks in the form of open water.

Introduction

Currently it is considered that one of the main factors of the Earth climate warming is the increased concentration of greenhouse gases in the atmosphere, including carbon dioxide (CO_2). Siberian ecosystems, comprising about 1/5 of the world's forested area, are the most

* This work was supported by grant of RSF №14-24-00113, as well as a grant from the RFBR №13-05-41506.

powerful reservoir of carbon and play an important role in the global carbon cycle affecting the Earth's climate [1].

Forest fires as an integral factor of forest ecosystem dynamics make a significant contribution to carbon emissions into the atmosphere. During the last few decades the incidence of fires in boreal forests has increased [2]. A number of studies forecast a further increase in fire frequency due to global climate change [3-5]. Thus, an important task is evaluation of changes in the atmosphere caused by CO2 emissions from forest fires.

This problem can be solved by integrating data from atmospheric research, ground-based observations and remote sensing. This work is part of research where joint analysis of following data is performed: state of the atmosphere obtained on the ZOTTO international observatory measuring mast; field studies for the mast coverage area; status of the underlying surface defined by satellite images.

The paper deals with the classification of satellite images to determine the main types of the underlying surface that have similar characteristics of the phytomass carbon stock, as well as analysis of long-term dynamics of the forest ecosystem status for the study area.

The study area is determined by the location of the ZOTTO international observatory (302 m height), which is installed near the Zotino settlement (Krasnoyarsk region territory, 60°48'N 89°21'E). Continuous monitoring of the greenhouse gas concentrations, radiation and meteorological parameters in the surface layers of the atmosphere are conducted at the measuring mast. [6]. The area of interest of this work is 100 km footprint of the ZOTTO. The study area is located in the middle taiga subzone of Western Siberia, are presented by the left and right banks of the Yenisei river. Major factors affecting forests in this area are wild fires and deforestation.

Classification of satellite images

Landsat ETM+ multispectral satellite imagery obtained during the 2000 vegetation period were used to identify the main types of the land surface. All Landsat images passed standard pre-processing including radiometric and geometric correction. Classification of satellite images for mapping land cover types is one of the main applications of remote sensing data [7]. Joint use of satellite and ground-based data improves the classification accuracy and allows to more accurately determining the spatial location of land cover types [8].

For Landsat imagery supervised classification by the maximum likelihood method was made. The maximum likelihood method is the most commonly used and most accurate for mapping the underlying surface [9]. As training sample for supervised classification were used ground data collected during field work in the years 2000-2014. These data consist of 350 permanent plots located in the most representative sites of forest stands. For these plots using the methods of enumerative inventory were specified characteristics of forest stands as well as based on these data body wood and wood-waste carbon content was calculated. All ground-based data is stored in the information system Forest, consisting of DBMS, GIS and the web server, enable remote data access at http://forest.sfu-kras.ru [10, 11].

The total area of the study is 3140 hectares. As a result of satellite data analysis for the ZOTTO footprint were specified 10 aggregated classes of land surface. Thematic map of land cover of the study area was obtained based on these classes. It was found that the forested area is about 76% of the study area. In 2000, a dominant position on the investigated territory is occupied by dark coniferous forests (spruce, fir, cedar forests) – 48 %. Deciduous forests (birch and aspen), formed on the felling and burned areas cover 12 % of the territory. About 16% of the study area is the pine biogeocenosis, represented by lichen and moss pine forests.

Wetland ecosystems occupy 11%, recent felling and scars occupy 3 and 5 % of the territory, respectively.

Monitoring of forest ecosystems

Satellite data of medium spatial resolution are widely used for forest ecosystems monitoring [12, 13]. In our work to study long-term changes in vegetation cover, we used Landsat ETM+, TM and OLI images. More than 20 scenes obtained from Landsat series satellites in the period from 2000 to 2014 were used. Distinctive feature of the Landsat program is the inheritance of spectral bands that provides continuity and compatibility of data obtained by different satellites of the Landsat series during the entire period of the program.

Identification of the felling and fire scars based on the values of spectral brightness of the Landsat images. As it is known, the change in the physiological status of vegetation leads to changes in their spectral characteristics [14]. This is achieved by reducing the absorption of sunlight by chlorophyll, and as a result, an increase in reflectance in the visible region of the electromagnetic spectrum. As well leaf tissue damage leads to lower reflectance in the near infrared region [15, 16]. These properties of vegetation are used for interpretation of satellite images.

When identifying forest areas damaged by fires, used the following criteria: spectral characteristics in the visible and infrared bands on the post-fire Landsat image; spatial features (presence of sharp boundaries); the presence of active fire polygons (hotspots) in the archive of the Krasnoyarsk center of space monitoring. The polygons of the active fires are calculated daily during the fire season by Terra,Aqua/Modis satellite data. For the detection of active fires threshold algorithm of Kaufman, based on the specifics of the fire radiation in the middle and far infrared range is used [17].

Results

Table 1 presents data on the number and area of fires detected in the study area over the years 2000-2014. As can be seen from table 1, the total area damaged by fire during the period of study is 852 560,28 ha. Some areas were affected by fire more than once. After excluding this areas the total area damaged by fire decreased to 830 037,61 ha. Thus, over the past 15 years, the fire damaged about 26% of the study area.

Table 1. The number and total area of fires in the years 2000-2014

Year	Area, ha	Number of fires
2000	8 025,12	3
2001	702,00	1
2002	0	-
2003	2 133,83	3
2004	6 606,16	8
2005	1 610,27	4
2006	73 412,47	22
2007	467,50	2
2008	653,60	1
2009	0	-
2010	0	-
2011	18 271,99	18
2012	711 063,53	33

2013	23 708,90	8
2014	5 904,91	9
Total for 2000-2014	852 560,28	112

As shown in table 1, two peaks of fire activity during the study period were observed, in 2006 and 2012. During the 2006 fire season more than 2% of the study area was damaged by fires, and 23% was damaged in 2012 fire season. Fires occurring in the last 15 years have made a significant contribution to the structure of forests and have led to an appreciable change of space occupied by vegetation ecosystems at the study area.

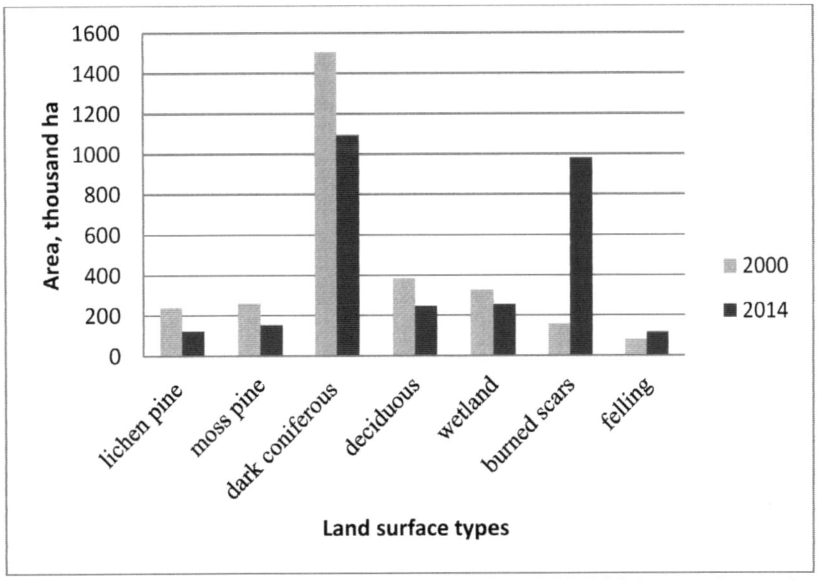

Figure 1. Changes in the ecosystem structure in 2000-2014 over the study area

As can be seen from figure 1, the area of burned scars for the investigation territory significantly increased in the period from 2000 to 2014 and has reached 980 thousand hectares. The area of felling has increased from 82 to 117 thousand hectares. Area of all types of vegetation ecosystems have declined: for lichen pine forests from 239 to 123 thousand hectares; for moss pine forests from 260 to 153 thousand hectares; for dark coniferous from 1502 to 1094 thousand hectares; for deciduous from 386 to 248 thousand hectares; for wetland ecosystems from 328 to 255 thousand hectares. Maximum numerical reduction is observed for dark coniferous, are dominant for this territory.

Conclusions

Thus, operational satellite monitoring of the investigated territory enables to assess the dynamics of forest disturbance by fire and logging over a longtime period and to estimate changes in forest ecosystems caused by these disturbance. In the course of the work were defined 10 significant classes of the underlying surface within ZOTTO 100 km footprint. Thematic map of the study area was composed by means of Landsat images supervised classification with the help of field data. Based on satellite data, it was found that the area of fire scars over the study period (from 2000 to 2014) has increased significantly, and currently is about 30% of the territory. Industrial logging does not cause significant damage to forest ecosystems of this territory: the total area of the felling at the moment is about 4%. The reduction of the area of all vegetated ecosystems was detected: lichen pine forests had lost 49% of its original area; moss pine forests – 41%, deciduous forests – 36%, dark coniferous forests – 27% and wetland ecosystems – 22% of their original territory.

It is known that the burning of forests is influenced by a number of factors such as forest type, stand age, moisture, wind speed, etc. [18, 19]. The results obtained in our work indicate the highest occurrence of fires for lichen forest types within study region, probably due to their high natural fire danger, which is consistent with other studies. The least damage the fire caused to the wetland ecosystem due to high content of moisture and the presence of a large number of fire breaks in the form of open water.

References

[1] Isaev A.S., Korovin G.N. *et al* 1995 Environmental problems in the absorption of carbon dioxide through reforestation and afforestation in Russia (Moscow, CREP)

[2] Kasischke E.S., Christensen N.L., Stocks B.J. 1995 Fire, global warming, and the carbon balance of boreal forests. *Ecol. Appl.* **5**, pp 437-451

[3] Krawchuk M.A., Cumming S.G., Flannigan M.D. 2009 Predicted changes in fire weather suggest increases in lightning fire initiation and future area burned in the mixedwood boreal forest *Climatic Change*. Vol **92**, pp 83-97

[4] Flannigan M.D., Amiro B.D. *et al* 2005 Forest fires and climate change in the 21st century *Mitigation and Adaptation Strategies for Global Change* **11(4),** pp 847-859

[5] Furyaev V. V. *et al* 2001 Effects of Fire and Climate on Successions and Structural Changes in the Siberian Boreal Forest *Eur. J. Forest Res* **2**, pp 1 -15

[6] Heimann M., Schulze E. D. *et al* 2014 The Zotino Tall Tower Observatory (Zotto): Quantifying Large Scale Biogeochemical Changes in Central Siberia *Nova Acta Leopoldina NF* **117(399)**, pp 51-64

[7] Zhuang Y. H., Liu X. J. *et al* 2013 Global Remote Sensing Research Trends during 1991–2010: A Bibliometric Analysis *Scientometrics* **96** pp 203–219

[8] Justice C., Belward A. *et al* 2000 Developments in the validation of satellite sensor products for the study of the land surface *Int. J. of Remote Sensing* Vol **21**, pp 3383 – 3390

[9] Yu L., Liang L. *et al* 2014 Meta-discoveries from a synthesis of satellite-based land-cover mapping research *Int. J. of Remote Sensing* Vol **35**, pp 4573-4588

[10] Antamoshkina O.A., Trofimova N.V., Verkhovets S.V. 2014 *The certificate of state registration of computer program* № 2014618974,.

[11] Antamoshkina O.A., Trofimova N.V. *et al* 2014 *The certificate of state registration of database* №2014620947,.

[12] Isaev A.S. *et al* 2014 Satellite remote sensing is a unique tool for monitoring of Russian forests *Vestnik Rossijskoj akademii nauk..* Vol **84,** 12 p 1073

[13] Kim D., Sexton J *et al* 2014 Global, Landsat-based forest-cover change from 1990 to 2000 *Rem. Sensing of Environment* **Vol 155**, pp 178-193

[14] Dejvis S., Landgrebe D., Filips T. 1983 Remote sensing: the quantitative approach (Moscow, Nedra)

[15] Kronberg P. Remote study of the Earth (Moscow, Mir)

[16] White J. D., Ryan K. C., Key C. C., Running S. W. 1996 Remote sensing of forest fire severity and vegetation recovery *Int. J. of Wildland Fire* **6** pp 125– 136

[17] Kaufman Y., Justice C. 1998 Algorithm Technical Background Document MODIS FIRE PRODUCTS

[18] Volokitina A.V., Sofronov M.A. 1996 Classification of plant combustible materials *Lesovedenie* **3** pp 38-44.

[19] Kurbatsky N.P. 1964 Fires of taiga, the regularities of their origin and development *Abstr. Dr.agr. sci. diss.* (Krasnoyarsk, FI)

IOP Conf. Series: Materials Science and Engineering **122** (2016) 012005 doi:10.1088/1757-899X/122/1/012005

The integral indicator of socio-economic assessment in regard to resource-oriented territories development in Russia

M N Chuvashova, N T Avramchikova, P V Zelenkov, M O Petrosyan
Reshetnev State University, 31 Krasnoyarsky Rabochy Av., Krasnoyarsk, 660037, Russia

E-mail: chuvashova@sibsau.ru

Abstract. Economic peculiarity of Russian resource-oriented territories are based on a focal type of industrial complex, differentiation of economies within a principle of mining and processing of natural resources. To improve the economic condition and integrate into the world innovative process is essential to solve the problem of eliminating the prevalence of resourse focus in the industrial economic structure that could ensure the overcoming of the existing spacial dissociation and market mechanisms development in innovative promotion. The monitoring system, involving the integral indicator of socio-economic and territorial potential assessment, has suggested by the authors. The integral indicator could guarantee the objective evaluation of economic condition within a territory that is vital for the governmental authorities to design strategies providing the economic development of administrative territories.

The economic space of the majority of Russian administrative territories are characterized by uneven economic development and its weak diversification at the same time they have significant natural and resource potential, their potential innovative form serves as a fund of vacant opportunities, which are the subject of close examination. [1]. Economic growth in Russia is mainly formed to due to mining and processing industry, which provides over half of all the budget revenues of the country. Emerging of economy at the cost of primary sector determines extensive rather than intensive modification of the development of the country. High prices on world commodity markets in primary resources have made a tangible contribution to the maintenance of high rate in GDP growth [2].

Economic peculiarity of Russian resource based territories are based on a focal type of industrial complex, differentiation of economies within a principle of mining and processing of natural resources. The quantity of basic fuel and power sectors in resource-oriented territories are considerably higher rather than in other territories of the country, which determines the pattern of development and problems of these particular territories. The economic growth is mainly determined due to the primary industries.

In the context of globalization of economic processes, a significant number of Russian territories have to solve the problem of eliminating the prevalence of resource-oriented focus on the industrial economic structure in order to provide the complete integration into the world innovative process that could guarantee the overcoming of the existing spacial dissociation and market mechanisms development in innovative promotion [3]. Recently an innovative approach towards the economic space has received an increasing attention provided by the researchers. This is due to the rapid upsurge of "knowledge economy" and innovative factor

IOP Conf. Series: Materials Science and Engineering **122** (2016) 012005 doi:10.1088/1757-899X/122/1/012005

highlight among the economic development agents [4]. The innovative distribution within the territorial socio-economic system and economic condition improvement as a whole require:

- the generation of income transfer mechanisms from the products sales of extractive industries in favour of branches with stiff technological modes;

-regional budget revenue sharing into oil and gas and others, creation of territorial reserve funds formed up by the partial income transfer from product sales of extractive industry branches;

- setting priorities in the territorial innovative processes development;

- further urbanization processes development (agglomeration) contributing to the appearance of scientific and technical ideas and innovative elaborations in accordance with the worldwide experience, that stimulates the initiative growth and territorial responsibility with regard to the efficient increase of development;

- further transport and energy infrastructure development of the territory (traffic and delivery acceleration, connectivity and disposition) which assures the innovation of development [5].

The authors, using the data from the Federal State Statistics Service of the RF, have carried out the analysis to identify the condition of basic economic indexes in comparison with the territories of Siberian Federal District in case of Krasnoyarsk Territory so that to determine and methodize the factors, influencing the technological and innovative development of resource-oriented territories.

Table 1. The main innovation development indexes within Krasnoyarsk Territory as a resource-oriented territory in comparison with the territories of Siberian Federal District and the RF in 2014 (in current prices).

Index	The Russian Federation	Siberian Federal District	Krasnoyarsk Territory
Gross regional product, bn. rub.	49920,0	5147,4	1192,6
Shipped internal goods, works and services, bn. rub. - including innovative products and services	41372,4 3507,9	4537,3 151,4	1060,5 53,9
Economic density of population (Ec.d=economically active population/(divided by) S km^2. Unit of measure a person per km^2)	4	2	1
Total freight turnover mln. ton per km.	250054	18057	3447
Volume of permanent investments, bn. rub.	13255,5	1378,0	369,3
Share in the volume of innovative production, %: - in GDP (GRP) - output of internal production	7,0 9,2	2,9 3,3	4,5 5,1
Net financial result, mln. rub.	6853753	446292	189769

Conducted studies testify that the output production and innovative production sales growth, as well as the growth of permanent investments volume within the economy of the territory of resource kind is extensively dependent on the territory richness in terms of human resources and development of industrial infrastructure. The analysis results allow to conclude that there is a weak correlation between the innovation development indicators of the territory and the net financial result (the share of permanent investments volume towards the net financial result is lower than in the Siberian Federal District (SFD) as a whole), which indicates the lack of interest on the part of regional economic business - community in the renewal of fixed assets and their modernization. Therefore, it has concluded that the market mechanisms efficiency of innovation diffusion is insufficient in resource – oriented territories. Besides the economic density of population in the territory (the ratio of economically active population to the total area of the territory) is generally much lower in comparison with the same indicator in SFD and Russia, so the innovative processes has characterized by the distinct local nature. The research results have approved the necessity of the support system establishment in the territories of such kind by the public authorities, who are in charge of the innovation infrastructure development, ensuring the creation and market promotion of most recent and upgraded products and technologies that will undoubtedly improve the quality of economic space inside a resource-oriented territory. It is essential to have an objective view of the region competitiveness, learn and evaluate the socio-economic potential of the territory to create and promote the most recent or innovative and upgraded products and technologies to the market. In this regard, it has proposed to develop a diagnostic system, which comprises an integral indicator of the socio-economic assessment of the territory development.

It is vital to carry out the diagnostics of the economic space of the territory as a whole to assess the level of regional economic development and the possibility of high-quality and uniform distribution of innovation in the resource-oriented territory. The diagnostics will show how qualitative the regional economic space is, so the integral indicator of the regional socio-economic development, which could been applied not only to the administrative territories of such kind, is suggested to be used. To calculate the integral index it is necessary to highlight the main directions of its socio-economic development and form an array of objective indicators for each of them. The absence of linear connection between the indexes consider one of the conditions in the indicator selection. The correlation analysis method uses to examine the existence of linear connection between the indicators. This particular research suggests Spearmen's identification pairwise test to guarantee the efficient discovery of linear connection (Table 2).

Table 2. The Structural integral indicator of the socio-economic evaluation of regional development.

Highlighted aspects of socio-economic regional development	Indicators of integral index of the socio-economic assessment of regional development	The integral indicator of the economic development level of the resource-oriented territory
Economic	- natural $(K_{1.1})$ - industrial $(K_{1.2})$ - investment $(K_{1.3})$ - labour $K_{1.4})$ - financial $(K_{1.5})$	$I=\sum_{i=1}^{4}\sum_{j=1}^{J_i} a_{ij}K_{ij}$ on condition of
Innovative	- the level of technological state of economic branches $(K_{2.1})$ - innovative potential $(K_{2.2})$	$\sum_{i=1}^{4}\sum_{j=1}^{J_i} a_{ij} = 1$
Social	- demographic $(K_{3.1})$ - population income level $(K_{3.2})$ - social population security $(K_{3.3})$	$0 \le a_{ij} \le 1$, i=1,4 j=1, J
Budgetary	- local balanced budget$(K_{4.1})$ - the rate of grants, subventions and subsidies from the budget of superior level $(K_{4.2})$ - support of enterprise development $(K_{4.3})$	K_{ij} – general index of importance of each indicator in a group. a_{ij}–ratio meaning, i- direction, j- indicator.

The adequacy of proposed integral indicator of socio-economic assessment of regional development are specified by the slow change of its ratings in terms of time because each of the integral properties reflects the conditions that can enhance or degrade the quality of economic space of resource-oriented territory under the consideration of time interval.

This particular approach, evaluating the quality of economic space within resource-oriented territories, complements the existing characteristics of its assessment and permits to carry out the correlation of the level of selected socio-economic lines of development, which are difficult to compare due to the diversity of evaluation factors. The essential content of the integral indicator in terms of socio-economic evaluation of the territorial potential lies in the integrated assessment in regard to the economic independence of the territory and possible development prospects. The integral indicator application in regard to the level evaluation of economic and innovative growth within the territories of various kinds contributes to the level determination of regional field development. The obtained results could applied as a series of recommendations concerning the administrative influence in order to improve the economic space of the territory. The integral indicator evaluation results would be used by the public authorities to design the economic development strategies in accordance with their resource profile.

References

[1] Scopin A. U. 2007. *Regional economic and environmental management.* Moscow. Prospect Publ. pp. 35-47.
[2] Hugget P. 2005. *General geography: global synthesis.* Pearson Custom Publ. pp. 233-242.
[3] Avramchikova N. T., Chuvashova M. N. 2015. *Regional economic: theory and practice.* No. 28 (403).
[4] Chuvashova M N, Avramchikova N T, Antamoshkin A N, *IOP Conf. Series: Materials Science and Engineering* 94 (2015) 012017.
[5] Chuvashova M. N. 2014. *Vestnik SibGAU.* no. 5 (57).

IOP Conf. Series: Materials Science and Engineering **122** (2016) 012006 doi:10.1088/1757-899X/122/1/012006

Intelligent control of PV system on the basis of the fuzzy recurrent neuronet*

E A Engel[1], I V Kovalev[2], N E Engel[3]

[1] Associate Professor, First Russian doctoral degree in computer sciences, Katanov State University of Khakassia, Abakan, RF,

[2] Professor, Second Russian doctoral degree in computer sciences, Siberian State Aerospace University, Krasnoyarsk, RF,

[3] Student, Katanov State University of Khakassia, Abakan, RF

E-mail: Angel@khsu.ru

Abstract. This paper presents the fuzzy recurrent neuronet for PV system's control. Based on the PV system's state, the fuzzy recurrent neural net tracks the maximum power point under random perturbations. The validity and advantages of the proposed intelligent control of PV system are demonstrated by numerical simulations. The simulation results show that the proposed intelligent control of PV system achieves real-time control speed and competitive performance, as compared to a classical control scheme on the basis of the perturbation & observation algorithm.

Introduction

Solar energy is being recognized as one of the main renewable energy sources due to its essentially non-polluting inexhaustible nature. We consider a non-linear maximum power point tracking (MPPT) problem [1] from photovoltaic applications. There are many existed MPPT algorithms such as voltage feedback, perturbation and observation, linear approximation, incremental conductance, hill climbing, actual measurement, fuzzy control and so on [1]. The goal of MPPT algorithms is to achieve the condition (dP/dV = 0) to find the maximum power point of PV system. MPPT algorithms required parameters are only the voltage and current of PV system. However the PV array is non-linear system and works under uncertainty of the environment due to sudden variations in the solar irradiation levels. Therefore unstable dynamics must be confronted when designing control systems to track maximum power point quickly – in order to provide stability, disturbance attenuation. Moreover, the non-linear photovoltaic (PV) system should be robust to different environmental conditions, in order to reliably generate maximum power. Therefore, automatic design methods utilizing intelligent techniques such as neural network and fuzzy logic are a promising alternative [2-6]. This forms the motivation for development of the effective control of PV system on the basis of the fuzzy recurrent neuronet presented in this paper. For real-life photovoltaic applications, the PV system behavior can change; system parameters can exhibit random variations; the solar irradiation can fluctuate. Hence neural-network based solutions have been proposed to overcome these difficulties. But the network needs to become more adaptive. Adaptive behavior can be enabled by modifying the network to have fuzzy and neural units which respond to the changing behavior of the photovoltaic system, thus making the network into a fuzzy recurrent neuronet. Therefore, photovoltaic systems are best modeled as hybrid systems since their behaviors result from the interaction between continuous and discrete dynamics. Several methods have been developed to overcome the aforementioned difficulties. The most important approaches are those that combine the basic MPPT algorithm, with the intelligent models, which are neural network and fuzzy logic. This paper presents the intelligent control of PV system on the basis of the fuzzy recurrent neuronet.

*This work is supported by Russian Foundation for Basic Research (grant №16-48-190912)

1. Control model based on perturbation & observation algorithm

One of the most common MPPT algorithm is perturbation & observation algorithm due to its simplicity, applicability and less required parameters. The main disadvantages of the perturbation & observation algorithm are: (1) the power tracked by the perturbation & observation algorithm can take considerable time to settle down from oscillating near the maximum power point; (2) if solar irradiation levels change suddenly then misjudgment phenomenon for the perturbation & observation algorithm takes place. Figure 1 shows such situation.

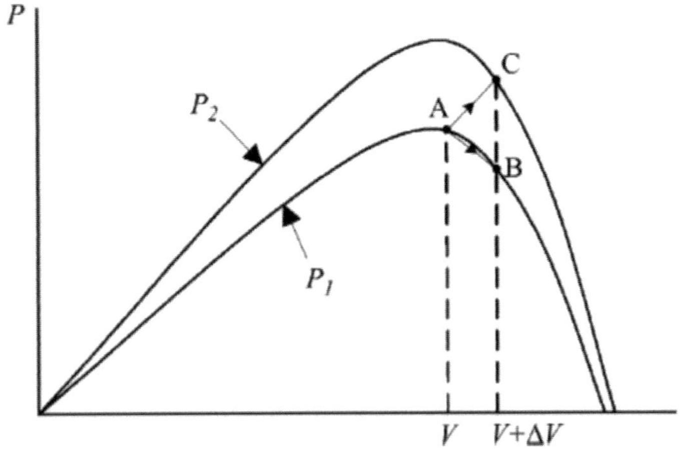

Figure 1. The misjudgment phenomenon for the perturbation & observation algorithm

The first point is point A, and ΔV voltage perturbation will move the operating point from A to B and will lead to decreasing power when the solar irradiation is constant. If the solar irradiation level increases in one sampling period, then the power curve will be moved from P1 to P2, and the operating point will be moved from A to C instead of A to B. Consequently, the power will be increased continuously and the voltage perturbation still moves toward ΔV direction. It can be seen that a continuous perturbation in one direction will lead to an operating point far away from the maximum power point. If the solar irradiation level increases without intermission, the distance between operating point and maximum power point will be farther. As a result, the power loss of photovoltaic system will increase, and the effectiveness of the photovoltaic system will reduce. In order to overcome the aforementioned difficulties the fuzzy recurrent neural net is proposed to track maximum power point of PV system.

2. Intelligent control of PV system on the basis of the fuzzy recurrent neuronet

In this paper, the function approximation capabilities of fuzzy recurrent neural net are used to approximate a nonlinear control law of maximum power point tracking. This paper considers the development of an effective control method that remains easy to implement. The proposed fuzzy recurrent neural net is capable of handling uncertainties in both the PV system parameters and the environment.

As formed, the fuzzy recurrent neural net creates the effective control signal, and identifies the system's state under sudden variations in the solar irradiation levels (Fig. 2). On the basis of the analysis of perturbation & observation algorithm for the adaptive maximum power point tracking it is necessary to trace voltage increment and solar irradiation changing. To make the intelligent control of PV system on the basis of the fuzzy recurrent neuronet become adaptive, it needs to have some idea on how the actual PV system's behavior is differing from its expected behavior, so that the fuzzy recurrent neural net can recalibrate its behavior intelligently during run time, and try to eliminate the constant maximum power point tracking error. Hence the input signal of the fuzzy recurrent neural net will be non-zero, and it will give useful feedback for telling how to adapt to the dynamically changing PV system's parameters. This control approach does provide a more intelligent method of implementing the control signal. The fuzzy recurrent neural net is trained based on the data

$$Z^i=(Ir^i, \Delta Ir^i, V^i, P^i, \Delta V^{i-2}, \Delta V^{i-1}, \Delta V^i), \quad (1)$$

where i – time, Ir is the solar irradiance, $\Delta Ir^i= Ir^i -Ir^{i-1}$, V represents the voltage, ΔV – the voltage increment, P – PV system's power; $x=(\Delta Ir^i, V^i, P^i, \Delta V^{i-2}, \Delta V^{i-1})$ – input signal of fuzzy recurrent neural net. The control signal and output signal of the fuzzy recurrent neural net is $u=\Delta V$. Data (1) have a training set of $9*10^5$ examples, and a test set of 10^5 examples. First, sets $AN_j^{\,s}(j\in\{1, ..., n^s\}, j=\overline{1..10})$ formed using GNG's (Growing Neural Gas [7]) clustering of the data (1); fuzzy sets $A_j^{\,s}(j\in\{1, ..., n^s\}, j=\overline{1..10}$ with membership function $\mu_j^s(x)$ formed using Fuzzy c-means of the data (1) (the sets $AN_j^{\,s}$ was input for Fuzzy c-means); in order to effective obtain the dynamics of the system the fuzzy sets A_j $(j\in\{1, 2, 3\})$ with membership function $\mu_j(x)$ determined according condition (2):

$$\eta=\frac{1}{n^s}\left(\sum_{u=1}^{n^s}\left[\min_{\substack{i=1\\ i\neq u}}^{n^s}[\mu_{A_u^s}(x),\mu_{A_i^s}(x)]\right]\right). \quad (2)$$

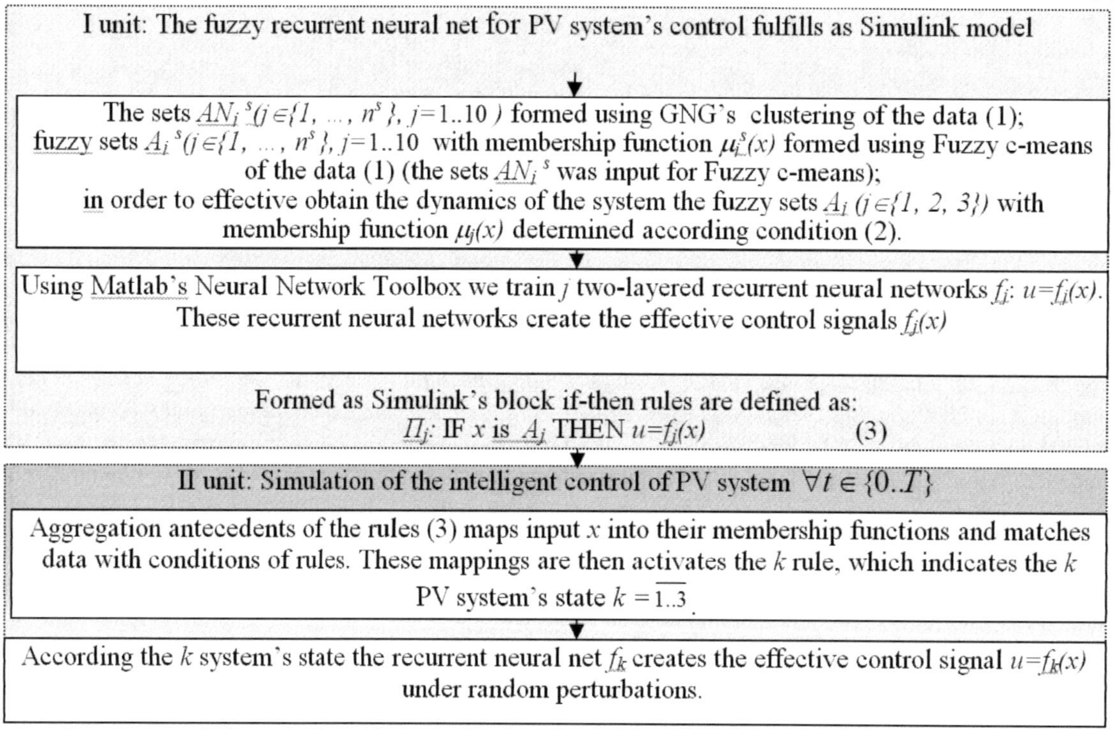

Fig. 2. The intelligent control of the PV system basis on the fuzzy recurrent neural net

Second, for each j an identifier is constructed by a two-layer feed forward neural network. Third, an fuzzy recurrent neural net for the photovoltaic system's control is carefully designed to correctly tackle the control task under uncertainty of the photovoltaic system and of the environment.

3. Matlab Simulink model of the photovoltaic system

The model of PV system contains:

- Two PV generators providing each a maximum of 0.1 MW at 1 kW/m^2 sun irradiance;
- 5-kHz boost converter;
- 20-kvar capacitor bank.

The model of photovoltaic system has two Kyocera-DD205GX-LP modules. This model uses a MPPT subsystem with a duty cycle that generates the required voltage to extract maximum power. To illustrate the benefits of the newly proposed intelligent control of the photovoltaic system's control, the numerical examples from the previous section 2 is revisited.

4. Simulation and results

All the simulations for this study are implemented in MATLAB, Simulink. The true benefits of the proposed the intelligent control of PV system on the basis of the fuzzy recurrent neuronet are best demonstrated through a simulation study. Figure 2 and figure 3 show the solar irradiance during simulation time for the first PV generator – PV1 and for the second PV generator -PV2 respectively.

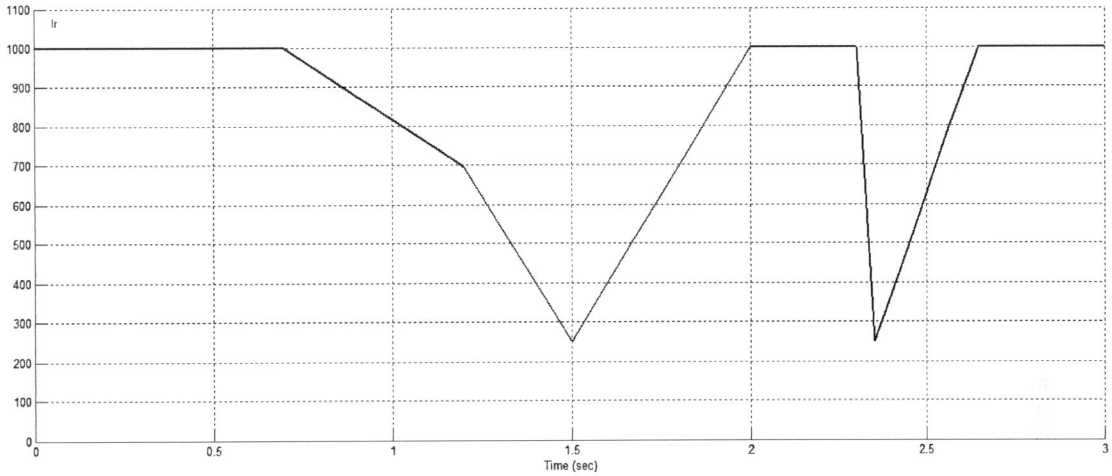

Figure 2. Plot of solar irradiance for PV1

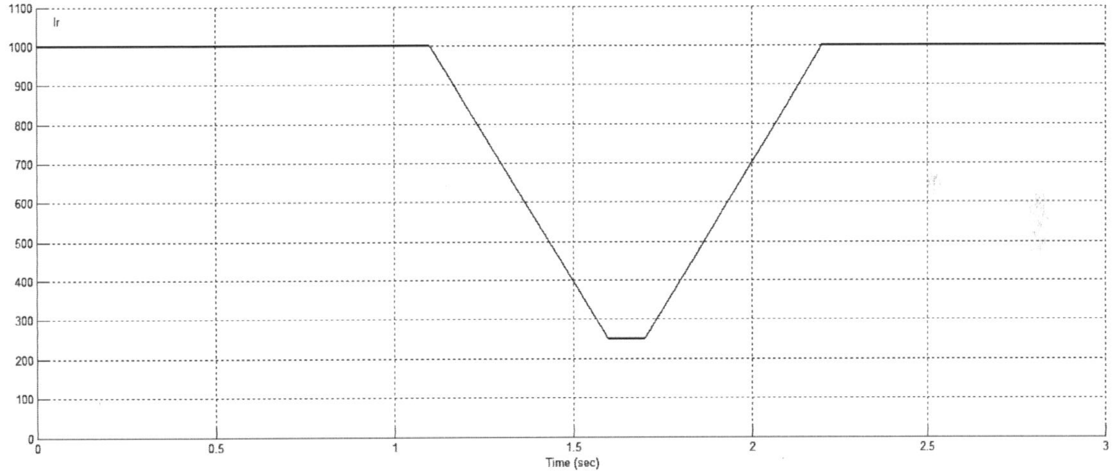

Figure 3. Plot of solar irradiance for PV2

In this comparison study, the performance of the energy saving technology of the PV system's control is compared against the standard model on the basis of the perturbation & observation algorithm, under the same conditions. Figures 4 to 6 illustrate the simulation's results. Figure 5 shows the misjudgment phenomenon for the perturbation & observation algorithm (black curve from time=2 s to 2.3 s). The intelligent control of PV system on the basis of the fuzzy recurrent neuronet is more robust and provided more power (Fig. 4-6) in comparison with the control model based on the perturbation & observation algorithm.

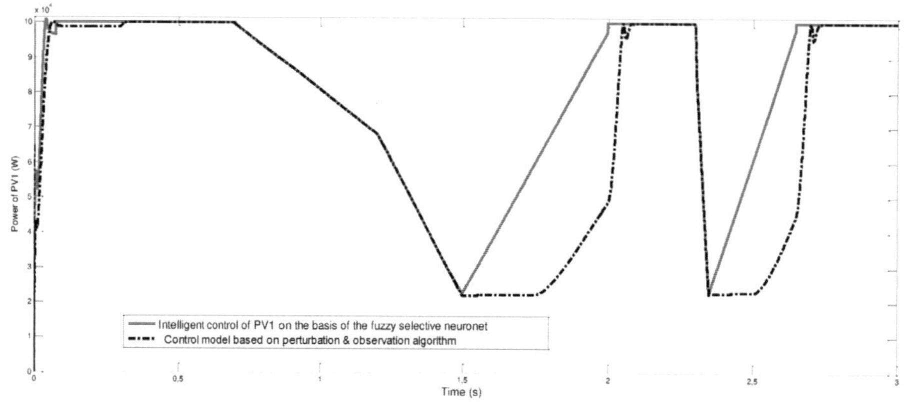

Figure 4. Plot of PV1's power provided by control model based on perturbation & observation algorithm and the intelligent control of PV1 on the basis of the fuzzy recurrent neuronet respectively

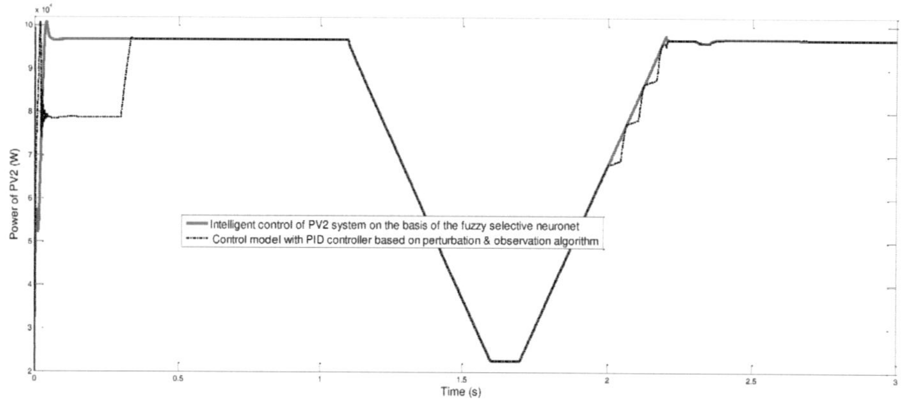

Figure 5. Plot of PV2's power provided by control model based on perturbation & observation algorithm and the intelligent control of PV2 array on the basis of the fuzzy recurrent neuronet respectively

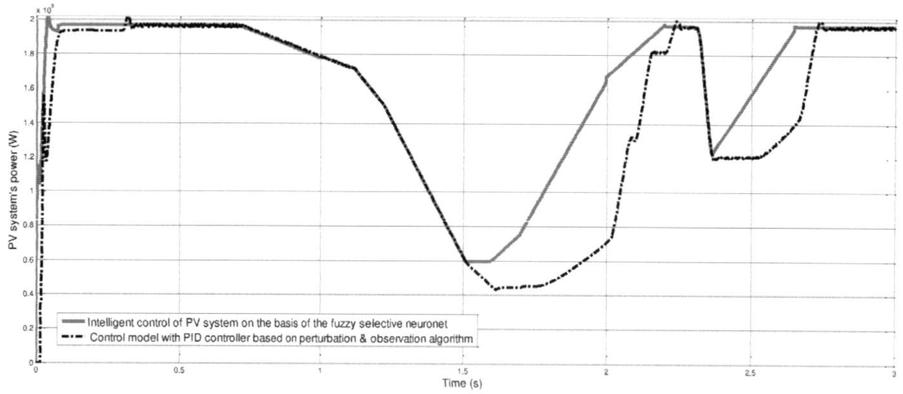

Figure 6. Plot of PV system's power provided by control model based on perturbation & observation algorithm and the intelligent control of PV system on the basis of the fuzzy recurrent neuronet respectively

The tolerance of a fuzzy recurrent neural net to noise is explained by two factors: the ability of the model to have similar responses for patterns contaminated with different intensities of noise and the resilience to noise of the low similarity of the responses for patterns of different PV system's state defined by sudden variations in the solar irradiation levels.

The use of the fuzzy recurrent neural net provides a more suitable approach to the MPPT problem, with the pointing accuracy. Extensive simulation studies on Simulink model have been carried out on different initial conditions, different disturbance profiles and variation in PV system and the solar irradiation levels parameters. It shows consistent performance has been achieved for the proposed intelligent control of PV system on the basis of the fuzzy recurrent neuronet with good stability and robustness as compared with the standard model based on perturbation & observation algorithm.

5. Conclusions

Extensive simulation studies show that the intelligent control of PV system on the basis of the fuzzy recurrent neuronet is robust to PV system uncertainties. Unlike popular perturbation & observation MPPT algorithm, a fuzzy recurrent neural net is used to approximate the control law, and not the system nonlinearities, which makes it suitable to handle a wide range of nonlinearities. Compared to standard MPPT algorithm – perturbation & observation, the intelligent control of PV system on the basis of the fuzzy recurrent neuronet produces good response time, low overshoot, and, in general, good performance. Simulation comparison results for a PV's control system demonstrate the effectiveness of the intelligent control of PV system on the basis of the fuzzy recurrent neuronet as compared with the standard model based on perturbation & observation algorithm.

References

[1] C. A. P. Tavares, K. T. F. Leite, W. I. Suemitsu, M. D. Bellar 2009 Performance evaluation of photovoltaic solar system with different MPPT methods // Industrial Electronics IECON '09. 35th Annual Conference of IEEE, pp.719-724

[2] Makarov I.M., Lokhin V. M., Manko S. V., Romanov M. P., Sitnikov M. S. 2013 Stability of intelligent systems of automatic control // Information Technology, 2

[3] Kovalev, I., Zelenkov, P., Ognerubov, S., Bahmareva, K., Denisova, E. 2015 The efficiency analysis of the automated plants // *IOP Conference Series: Materials Science and Engineering* 70 (1)

[4] Engel, E. , Kovalev, I.V. , Kobezhicov, V. 2015 Intelligent control of non-linear dynamical system based on the adaptive neurocontroller // IOP Conference Series: Materials Science and Engineering 94 (1)

[5] Engel, E. , Kovalev, I.V., D Karandeev 2015 Energy-saving technology of vector controlled induction motor based on the adaptive neuro-controller // IOP Conference Series: Materials Science and Engineering 94 (1)

[6] Engel E.A. 2012 Solving control problems, decision making and information processing by fuzzy selective neural network// Information Technology, 2

[7] Isaac J. Sledge 2008 Growing Neural Gas for Temporal Clustering/IEEE

IOP Conf. Series: Materials Science and Engineering **122** (2016) 012007 doi:10.1088/1757-899X/122/1/012007

Vector control of wind turbine on the basis of the fuzzy selective neural net *

E A Engel[1], I V Kovalev[2], N E Engel[3]

[1] Associate Professor, First Russian doctoral degree in computer sciences, Katanov State University of Khakassia, Abakan, RF,

[2] Professor, Second Russian doctoral degree in computer sciences, Siberian State Aerospace University, Krasnoyarsk, RF,

[3] Student, Katanov State University of Khakassia, Abakan, RF

E-mail: Angel@khsu.ru

Abstract. An article describes vector control of wind turbine based on fuzzy selective neural net. Based on the wind turbine system's state, the fuzzy selective neural net tracks an maximum power point under random perturbations. Numerical simulations are accomplished to clarify the applicability and advantages of the proposed vector wind turbine's control on the basis of the fuzzy selective neuronet. The simulation results show that the proposed intelligent control of wind turbine achieves real-time control speed and competitive performance, as compared to a classical control model with PID controllers based on traditional maximum torque control strategy.

Introduction

At the present time, because of ecological problems and due to limitation of natural energy sources the renewable energy becomes more popular. The wind energy is the most developing technology of the alternative clean energy sources. Modern wind turbine stations and windmill farms compiles the complex system for electricity generation. In recent years, a number of advanced control schemes have been proposed, but most of them lack the capacity to address the multi-objective problem since they concentrate only on particular objectives while disregarding others [1]. However, the controller did not take into account the fluctuation of structural loads. In this paper three–phase Permanent Magnet Synchronous Generator (PMSG) that will be used to generate electrical energy driven by wind turbine is simulated by Matlab.

There are many classical control strategies for wind turbine such as most common decoupling algorithm & maximum power point tracking (MPPT) and traditional maximum torque control strategy. One of the most common control strategies is the Proportional-Integral-Derivative (PID) controller due to its simplicity and applicability [2]. However, wind turbine systems are non-linear and commonly suffer from restrictions imposed by the uncertainty of the environment as a result of sudden variations in the wind speed. Therefore unstable dynamics must be confronted when designing control systems to track maximum power point quickly – in order to provide stability, disturbance attenuation. But the PID controllers for non-linear wind turbine system have slow response times to changing reference commands, take considerable time to settle down from oscillating around the MPP state, must often be designed by hand, and require extensive analysis of the system and dynamics. This process is generally difficult because it is hard to anticipate all operating conditions. The control model must coordinate the non-linear wind turbine system properly, generating robust behavior to negotiate disturbance attenuation effectively while maintaining stability. Moreover, the non-linear wind turbine system should be robust to different environmental conditions, in order to reliably generate maximum power.

Therefore, automatic design methods utilizing intelligent techniques such as neural network and fuzzy logic are a promising alternative [2-6]. For real-life wind turbine applications, the wind turbine system behavior can change; system parameters can exhibit random variations; the wind speed can fluctuate. Hence neural-network based solutions have been proposed to overcome these difficulties. But the network needs to become more adaptive.

*This work is supported by Russian Foundation for Basic Research (grant №16-48-190912)

Adaptive behavior can be enabled by modifying the network to have fuzzy units which respond to the changing behavior of the wind turbine system. Therefore, wind turbines are best modeled as hybrid systems since their behaviors result from the interaction between continuous and discrete dynamics. Several methods have been developed to overcome the aforementioned difficulties. The most important approaches are those that combine the classical control strategy with the intelligent models, which are neural network and fuzzy logic. This paper presents the vector control of PV system on the basis of the fuzzy selective neuronet.

1. Wind turbine control model based on PID controllers

The aim of the Field Oriented Control is to perform real time control of the torque demand, to control rotor mechanical speed and to regulate phase currents. To perform these controls, the electrical equations are projected from a 3 phase non-rotating frame into a two co-ordinate rotating frame [1].

Using Clarke –Park transformations voltage and torque equations of the PMSG in d-q rotor reference frame can be given by (18)-(22), where the q-axis goes ahead 90 degrees from the d-axis with respect to the direction of rotation.

$$u_{sd} = R_s i_{sd} + \frac{d}{dt}\Psi_{sd} - w_r \Psi_{sq} \tag{18}$$

where Ψ_{sd} is the d-axis stator flux linkage, Ψ_{sq} is the q-axis stator flux linkage.

$$u_{sq} = R_s i_{sq} + \frac{d}{dt}\Psi_{sq} + w_r \Psi_{sd} \tag{19}$$

$$\Psi_{sd} = L_s i_{sd} + \Psi_m \tag{20}$$

$$\Psi_{sq} = L_s i_{sq} \tag{21}$$

$$T_e = \frac{3}{2} p(\Psi_{sd} i_{sq} - \Psi_{sq} i_{sd}) \tag{22}$$

where u_{sd} and u_{sq} are voltages in d-q axis , R_s is the stator resistance, L_s is the stator inductance, w_r is the electrical rotational speed, Ψ_m is the permanent magnet flux constant provided by the magnets, and p is the number of pole pairs. In this paper we ignore losses and can write equation for the dc-link dynamics as follows:

$$i_{gen} = i_c + i_{grid} \tag{23}$$

$$i_{gen} = \frac{P_{gen}}{v_c} \tag{24}$$

$$P_{gen} = \frac{3}{2}(u_{ds} i_{ds} + u_{qs} i_{qs}) \tag{25}$$

$$i_c = C\frac{dv_c}{dt} \tag{26}$$

where P_{gen} is the generator output power, i_{gen} is the generator output current flowing towards the capacitor bank and the grid side, i_{grid} is the current drawn by the grid side from the dc bus, v_c is the capacitor voltage across the DC-link.

Vector control used a field-oriented theory controls space vectors of magnetic flux, current, and voltage. A traditional maximum torque control strategy has the rotor flux oriented frame. In this frame the q-component reference of the stator current is generated from a speed controller while the d-component reference is set to zero.

Vector control presents following benefits: speed control over a wide range, precise speed regulation, fast dynamic response, and operation above base speed.

2. Vector control of wind turbine system on the basis of the fuzzy selective neuronet

In this paper, the function approximation capabilities of fuzzy selective neural net are exploited to approximate a nonlinear control law of maximum power point tracking. The proposed fuzzy selective neural net is capable of handling uncertainties in both the wind turbine system parameters and the environment.

As formed, the fuzzy selective neural net creates the effective control signal, and identifies the system's state. To make the vector control of wind turbine system on the basis of the fuzzy selective neuronet become adaptive, it needs to have some idea on how the actual wind turbine system's behavior is differing from its expected behavior, so that the fuzzy selective neural net can recalibrate its behavior intelligently during run time, and try to eliminate the constant maximum power point tracking error. Hence the input signal of the fuzzy selective neural net will be non-zero, and it will give useful feedback for telling the model how to adapt to the dynamically changing wind turbine system's conditions.

In this research we used experiment's data $Z^i=(X^i, y^i=(u_1^i, u_2^i))$, where $i=\overline{1..10^6}$, $X\in R^{21}$ – represent the wind turbine's input and some output data, stator current torque – $i_{sa}=u_1$ and flux $-i_{sd}=u_2$. We did dimensionality reduction of data $X\in R^{21}$ from 21 to 5. This dimensionality reduction is developed through Principal Component Analysis. This step forms training data set (x, i_q) for the selective neural network ($x \subset X$, $x\in R^5$). The statistical features $x=(i_{sa}, i_{sb}, i_{sc}, \Delta P_{gen}=P_{gen}^{i}-P_{gen}^{i-1}$, electromagnetic moment, speed of rotor) extracted from the data Z are used as inputs to fuzzy selective neural network. The selective neural net is trained based on the data

$$z^i=(x^i, y^i) \tag{1}$$

where $i=\overline{1..10^6}$, x – input signal of fuzzy selective neural net; y – control signal and output signal of fuzzy selective neural net.

First, sets $AN_j^{\,s}(j\in\{1, ..., n^s\}, j=\overline{1..10})$ formed using GNG's (Growing Neural Gas [7]) clustering of the data (1); then, fuzzy sets $A_j^{\,s}(j\in\{1, ..., n^s\}, j=\overline{1..10}$ with membership function $\mu_j^s(x)$ formed using Fuzzy c-means of the data (1) (the sets $AN_j^{\,s}$ was input for Fuzzy c-means); in order to effective obtain the wind turbine's dynamic the fuzzy sets A_j $(j\in\{1, 2, 3\})$ with membership function $\mu_j(x)$ determined according condition (2):

$$\eta=\frac{1}{n^S}\left(\sum_{u=1}^{n^s}\left[\sum_{\substack{i=1\\ i\neq u}}^{n^s}\min[\mu_{A_u^s}(x), \mu_{A_i^s}(x)]\right]\right). \tag{2}$$

Second, for each j an identifier is constructed by a two-layer feed forward neural network. The data (1) have a training set of $9*10^5$ examples, and a test set of 10^5 examples. Third, Formed as Simulink's block if-then rules are defined as:

$$\Pi_j\text{: IF } x \text{ is } A_j \text{ THEN } u=f_j(x). \tag{3}$$

Thus, the fuzzy selective neural net for the wind turbine's control was carefully designed to correctly tackle the control task under uncertainty of the wind turbine system and of the environment.

In this research the fuzzy selective neuronet is used for vector control of wind turbine. The vector wind turbine's control on the basis of the fuzzy selective neuronet elaborates following steps:

- Measure the generator quantities (phase voltages and currents),
- Transform the quantities into a stationary 2-phase (α,β) system using Clarke transformation,
- Calculate the rotor flux space vector magnitude and position angle, input x,
- Aggregation antecedents of the rules (3) maps input x into their membership functions and matches data with conditions of rules. These mappings are then activates the k rule, which indicates the k wind turbine's state $k=\overline{1..3}$.
- Transform stator currents into the d-q coordinate system using Park transformation,
- The stator current torque – $i_{sq}=u_1$ and flux $-i_{sd}=u_2$ producing components are created by neural net f_k: $(u_1, u_2)=f_k(x)$,
- The stator voltage space vector is transformed back from the d-q coordinate system into the two-phase system and fixed to the stator by inverse Park transformation.

This vector wind turbine's control on the basis of the fuzzy selective neuronet does provide a more intelligent method of implementing the control signal. The proposed scheme with the fuzzy selective neural was implemented using MATLAB/Simulink software.

3. Simulation and results

In this section, simulation studies are accomplished to clarify the applicability and benefits of the proposed vector wind turbine's control on the basis of the fuzzy selective neuronet. All the simulations for this study are implemented in MATLAB, Simulink. The simulation of the traditional maximum torque control strategy is obtained considering the expressions presented in sections 1. The simulation of the intelligent control of wind turbine is obtained fulfilling the fuzzy selective neuronet presented in sections 2. Boost converter and control algorithms are also implemented and included in Simulink models.

In this comparison study, the performance of the vector wind turbine's control on the basis of the fuzzy selective neuronet is compared against the standard control model with PID controllers based on traditional maximum torque control strategy, under the same conditions. Figures 1 to 2 show the simulation's results. Figure 1 shows that the intelligent control of wind turbine provided a reduction in the steady-state and transient wind turbine's rotor speed as compared with the standard control model with PID controllers.

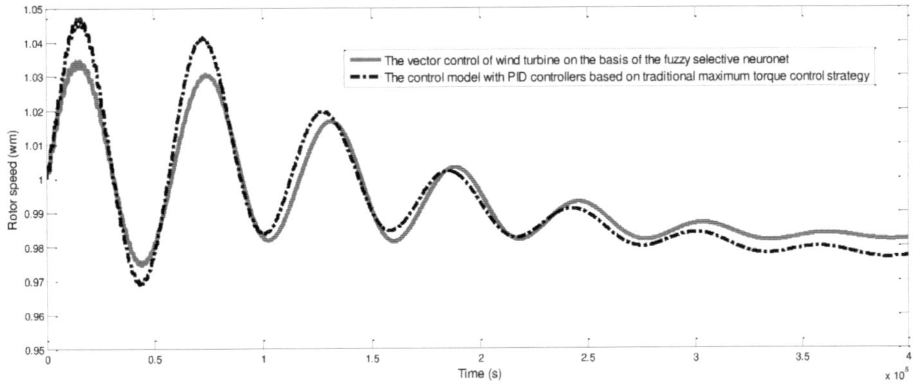

Figure 1. Plot of wind turbine's rotor speed provided by intelligent control of wind turbine and control model with PID controllers based on traditional maximum torque control strategy respectively

Figure 2 shows that the intelligent control of wind turbine provided a reduction in the steady-state and transient PMSG torque as compared with the standard control model with PID controllers. Figure 3 shows that the proposed intelligent control of wind turbine provides energy saving due reduction in the steady-state and transient PMSG torque compared with the control model with PID controllers based on traditional maximum torque control strategy.

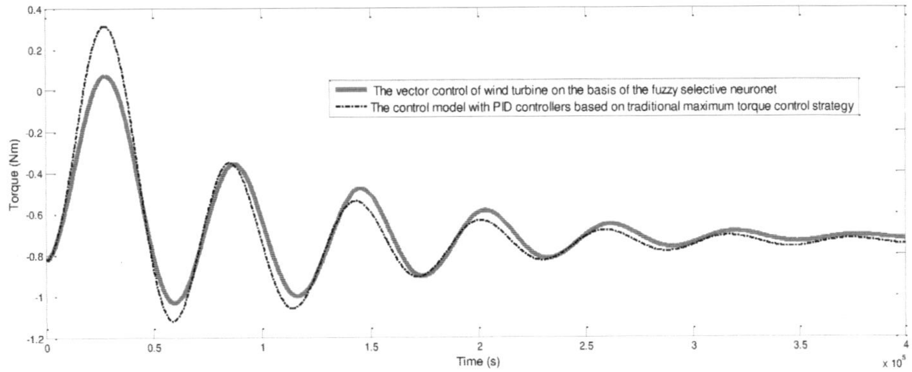

Figure 2. Plot of PMSG torque provided by intelligent control of wind turbine and control model with PID controllers based on traditional maximum torque control strategy respectively

IOP Conf. Series: Materials Science and Engineering **122** (2016) 012007 doi:10.1088/1757-899X/122/1/012007

The vector control of wind turbine on the basis of the fuzzy selective neuronet is more robust and provided more power (Fig. 3) in comparison with the control model with PID controllers based on traditional maximum torque control strategy.

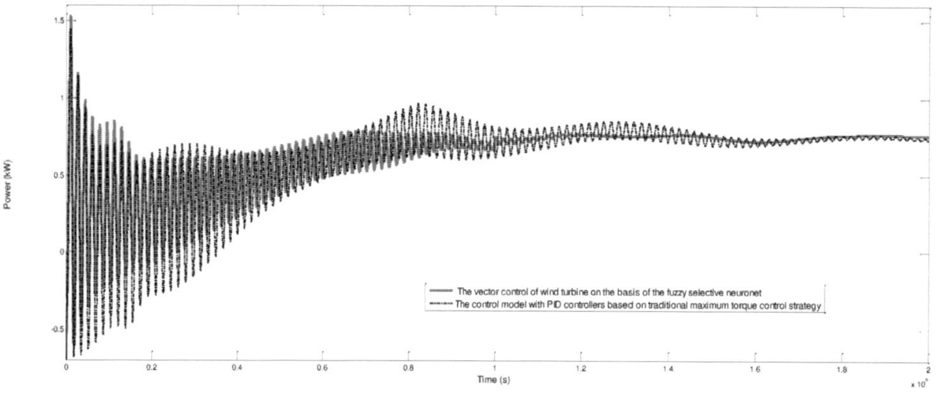

Figure 3. Plot of wind turbine's power provided by intelligent control of wind turbine and control model with PID controllers based on traditional maximum torque control strategy respectively

The aforementioned dynamic simulation results performance and robustness of the proposed vector control of wind turbine on the basis of the fuzzy selective neuronet are confirmed as compared to control model with PID controllers based on traditional maximum torque control strategy respectively.

The use of the fuzzy selective neural net provides a more suitable approach to the MPPT problem, with the pointing accuracy. Extensive simulation studies on Simulink model have been carried out on different initial conditions, different disturbance profiles and variation in the wind speed levels parameters. It shows consistent performance has been achieved for the proposed fuzzy selective neural net with good stability and robustness as compared with the standard control model with PID controllers based on traditional maximum torque control strategy.

4. Conclusions

It is shown that the vector control of wind turbine on the basis of the fuzzy selective neuronet is robust to the wind turbine's control system uncertainties. Unlike popular control model with PID controllers based on traditional maximum torque control strategy, a fuzzy selective neural net is used to approximate the control law, and not the system nonlinearities, which makes it suitable to handle a wide range of nonlinearities. Compared to standard control model with PID controllers based on traditional maximum torque control strategy, the vector control of wind turbine on the basis of the fuzzy selective neuronet produces MPPT, good response time, and performance. Simulation comparison results for the wind turbine's control system demonstrate the effectiveness of the vector control of wind turbine on the basis of the fuzzy selective neuronet as compared with standard control model with PID controllers based on traditional maximum torque control strategy.

References

[1] Xiaodong W., Shixu H., Shirong W., Yingming L., Lixia L. 2013 Multi objective optimization torque control of wind turbine based on LQG optimal control // Control and Decision Conference (CCDC), 405-408.

[2] Makarov I.M., Lokhin V. M., Manko S. V., Romanov M. P., Sitnikov M. S. 2013 Stability of intelligent systems of automatic control // Information Technology, 2

[3] Kovalev, I., Zelenkov, P., Ognerubov, S., Bahmareva, K., Denisova, E. 2015 The efficiency analysis of the automated plants // *IOP Conference Series: Materials Science and Engineering* 70 (1)

[4] Engel, E. , Kovalev, I.V. , Kobezhicov, V. 2015 Intelligent control of non-linear dynamical system based on the adaptive neurocontroller // IOP Conference Series: Materials Science and Engineering 94 (1)

[5] Engel, E. , Kovalev, I.V., D Karandeev 2015 Energy-saving technology of vector controlled induction motor based on the adaptive neuro-controller // IOP Conference Series: Materials Science and Engineering 94 (1)

[6] Engel E.A. 2012 Solving control problems, decision making and information processing by fuzzy selective neural network// Information Technology, 2

[7] Isaac J. Sledge 2008 Growing Neural Gas for Temporal Clustering/IEEE

IOP Conf. Series: Materials Science and Engineering **122** (2016) 012008 doi:10.1088/1757-899X/122/1/012008

Experience of e-learning implementation through massive open online courses

N V Ivleva and E V Fibikh
Business Foreign Languages Department
Reshetnev Siberian State Aerospace University,
31, Krasnoyarsky Rabochy Av., Krasnoyarsk, 660037, Russian Federation

E-mail: natalie_ivleva@mail.ru

Abstract. E-learning is considered to be one of the most prospective directions in education development worldwide. To have a competitive advantage over other institutions offering a wide variety of educational services it is important to introduce information and communication technologies into the educational process to develop e-learning on the whole. The aim of the research is to reveal problems which prevent from full implementation of e-learning at the Reshetnev Siberian State Aerospace University (SibSAU) and to suggest ways on solving those problems through optimization of e-learning introduction process at the university by motivating students and teaching staff to participate in massive open online courses and formation of tailored platforms with the view to arrange similar courses at the premises of the university. The paper considers the introduction and development level of e-learning in Russia and at SibSAU particularly. It substantiates necessity to accelerate e-learning introduction process at an aerospace university as a base for training of highly-qualified specialists in the area of aviation, machine building, physics, info-communication technologies and also in other scientific areas within which university training is carried out. The paper covers SibSAU's experience in e-learning implementation in the educational process through students and teaching staff participation in massive open online courses and mastering other up-to-date and trendy educational platforms and their usage in the educational process.

Key words. E-learning, distance learning, online learning, massive open online course.

1. Introduction

Term e-learning, which means the educational process by electronic means through the Internet facilities, became widespread during the last 2-3 years. The process of conversion from traditional education process to the computer-based one has been gradually developed during the last decades. Since the large, represented on the machine-readable carriers, archives appearance and the amount of information and knowledge enlargement, frequently appears the idea to use this material for educational purposes, as being the only form that is able to keep pace with intellectual content generation. In the occurred situation the e-learning stands for the new technological environment of knowledge delivery from the open sources. Globally, the e-learning became available due to the Internet development, that provides the opportunity to send the required quantity of data from one part of the world to another, communicate easily with other net users online and post information on Internet websites, making them available for anyone. Finally, the appearance of education through the information technologies is due to the market mechanisms influence. Growing demand for higher education resulted in the educational services market appearance, where universities became competing stakeholders in the educational content distant delivery area.

If we review the steps of e-learning by using computer technology development, then they could be represented in a following way:

- the courses on removable media;
- educational websites;
- electronic online platforms.

2. Implementation of e-learning at SibSAU

Nowadays the application of the distant learning based on Moodle platform actively used and enhanced [1]. This learning is applied for part-time and additional educational program's students. This educational process is based on motivated and controlled intensive self-dependent work of the learner, who can study in the convenient place due to the individual schedule, carrying special learning tools set and concerted opportunity to contact with the teacher. Talking about advantages of the distant learning, first of all, it has positive sides for students, like: learning schedule's flexibility, opportunity to study by individual plan according to their own needs and abilities, unprejudiced and independent from the professor's knowledge assessment system, opportunity to consult with the teacher during the study. Such form of education primarily means the appearance of additional possibility for the teachers to present material to students, i.e. de facto the opportunity to teach more students within the same teaching load appears.

Recently, involvement of the Internet in the education, quantity of online courses, their themes, various ways of realization and the general orientation on the whole ended up with more capacious term – e-learning. European committee defines e-learning as "the usage of modern media and Internet technologies for education quality increase at the expense of resources and services accessibility enhancement, and also distant knowledge exchange and joint work". Today e-learning is the educational process, where interactive electronic information delivery ways are used: removable media, corporate networks, the global network.

Despite achieving its first-priority goal as distant learning through the Internet, e-learning is also a great addition to the full-time study and can be a good support for traditional study's quality and effectiveness increase. Obvious advantages of e-learning are: vast freedom of access – the student has an access to the electronic courses through the Internet from any place, where global information network connection exists; competent, qualified education – the courses are created by the whole team of specialists, what makes e-learning mature and qualitative one; opportunity for electronic course content separation into modules, small blocks of information, that allows to make studying process more flexible and simplifies search of the necessary materials; flexibility of studying process – duration and order of learning material are chosen by the student himself, completely adapting the whole studying process to fit his abilities and needs; opportunity to study in the workplace – students can acquire education on-the-job (if there is any), at home and on the way using mobile Internet; opportunity to develop with the times – electronic courses users: both teachers and students develop their skills and knowledge due to the cutting-edge technologies and standards, electronic courses allow to promptly refresh educational materials; opportunity to define criteria for accessing knowledge – during e-learning there is opportunity to specify clear criteria of knowledge assessment, that the student gained during the educational process. Undoubtedly, advantages of e-learning are obvious: flexibility, time saving, ease to review what has already been learned [2].

Such form of education is also wide spread in SibSAU for full-time students' education, for instance, during foreign language courses.

3. Massive open online courses as e-learning development method

One of the e-learning elements applied at practice are Massive Open Online Course (MOOC). Massive open online course is the distant Internet-learning method that is remarkable for the following characteristics:

- a large number of courses participants as compared to regular university courses;
- despite commercial nature of MOOC business the course has open pattern of its structure, content and educational aims. However, it can have additional commercial components, for example, in terms of paid certificates issue at the end of the course;
- a course is held completely online with the help of both asynchronous and synchronous educational methods.

Wide range of informational and educational technologies is used in the process of MOOC's development and realization. Basic technologies are integrated video lectures with integrated presentations, graphics, interactive tasks and text information. These elements allow increasing the level of tasks interactivity that can hasten educational process and reveal problem blocks in studying process to focus on in future. Moreover, the course's participants motivation arise should be noted, they can see and realize the essence and importance of processes studied within the course, and can also view the result right after the task is done, analyze mistakes and pursue their own educational line of the further study inside of the course. The second important MOOC's technology is accessing each other. This technology enables to access the competence level of other course participants, who are, either way, united by the course theme. Because of this technology the interactive communication occurs not only between students and teachers, but also between students among themselves through the forums, communities and other integrated software. Provided opportunity increases participants' competitiveness and brings them to the more proficient level of knowledge. Other technologies, like wiki-projects, researching and team nature of work also play a vital role in MOOC.

Implementing such technologies, more attention is paid to advantages, rather than disadvantages, because they negligible.

One of the advantages is the fact that there is an opportunity to organize MOOC in any environment, where the connection is available, in any language, using different online instruments of the target area. The education process takes place in a more informal form that helps to overcome a psychological barrier (fear to make mistake or speak in a wrong way, fear to be disapproved by other project's participants or teachers, etc.) and gain knowledge freely. There is no need for participants to provide educational qualifications or other document, confirming their education level or identity. Motivation is the only thing that is needed. With the motivation participants enhance living skills needed for study. The participation in MOOC's program constrains to think about own education and knowledge acquiring.

Monitoring existing international Internet-platforms, on a base of which MOOC is realized, the following should be noticed: Coursera (www.coursera.org) [3], Futurelearn (www.futurelearn.com) [4], Cambridge English Teacher (www.cambridgeenglishteacher.org) [5], EdX (www.edx.org) [6], Udacity (www.udacity.com) [7], Khan Academy (www.khanacademy.org) [8].

SibSAU's professors have experience using and working with such a MOOC platforms as Coursera, Futurelearn and Cambridge English Teacher. It's worth noting that the first two platforms are used by both teachers and full-time students. The Cambridge English Teacher is used only by teachers from the *Business Foreign Languages and Technical Foreign Languages* Departments.

Coursera is the project in massive online education sphere, founded by Stanford computer science professors Andrew Ng and Daphne Koller. The project collaborates with the universities that publish and run courses in the system on different fields of knowledge. Students take the courses, communicate with fellow students, take tests and exams directly on the Coursera website, and also the mobile application for iPhone and Android is distributed. As at November 2014 10 million users and 844 courses from 108 educational institutions were registered in Coursera.

Futurelearn is the project offering a wide range of online courses, lectured by leading universities and culture institutes around the world. These courses are available not only from personal computers, but also from applications of mobile devices. Thus, course participants have an opportunity to study at any place, in any time if there is Internet access. The founder is British Open University, which has 40-years' experience in online education development. The project is supported by 40 partners from Great Britain, Europe, Africa, Asia and Middle East. The first course was open in September 2013. In a year approximately 500 000 participants joined the project.

Cambridge English Teacher is the project launched especially for English language teachers of the all educational levels (primary, secondary, higher education) by Cambridge University. The aim of this project is professional competency development in the methods of English language teaching for various purposes area. Like the other MOOC platforms Cambridge English Teacher offers participation in online courses oriented at the increase of both language and professional skills. However, the resource provides the opportunity to take part in online courses, but to attend webinars, forums as well, to get an access to the printed material in e-form that is essential help to the English language teachers in lessons development and conducting taking into account new tendencies in the international society.

The university as being high education institute have to go with the times, tracing all new tendencies in education, applying them for the highly qualified specialists training for organizations and enterprises. MOOC help to promote the university brand in the international education services markets, attracting talented youngsters. Such promotion is one of the steps of accomplishing the mission set by the government to universities, namely, to reinforce the export of Russian educational services. Participating in the projects of MOOC realization, the university gets additional financing through cofounding. This advantage provides the opportunity to the university to receive additional income apart from budgetary funds.

Because of the fact that SibSAU is a regional university, located far away from leading country's universities, the participation in MOOC projects provides an opportunity of joint network usage of the resources with large universities, and to make up the shortage of professors of not enough developed educational directions. Taking this into account, the opportunity of collaboration appears not only with teachers, but also with specialists and experts in the areas where strong connection with employers, business structures and outer professional world is needed. The collaboration of this kind will increase the university's reputation, will provide an opportunity for international cooperation development, and also invited specialists can contribute to university's projects development and taking them on the international significance level.

Despite MOOC technologies just started to be implemented in Russian universities, the successful tendency of development of this direction in Russia can already be traced. The work is being conducted in the direction of common institutional online departments, virtual scientific laboratories formation in Russian universities; creation of intercollegiate and inter-institutional scientific and educational programs; science and education integration. The global goal that is pursued on the federal level is the creation of Federal electronic researching university.

According to the results of the research, conducted by Education Development Centre of Moscow School of Management Skolkovo, the online education in Russia develops for a long time, but there are already no well-known projects, unlike European countries and USA. As the researchers suggest, this may be connected with the fact that almost all existing Russian online platforms were initially directed at paid rendering educational services and, consequently, are not pretending to cover a large audience [9]. Considering the fact that the government is interested in MOOC technologies realization in Russia, as they can be a network education variant, which is the top-priority aim of development of the education in the country, than the possibility of leading Russian universities motivation for rendering the free access to their resources cannot be excluded [10].

Among Russian universities collaborating with projects in the field of online education, the most active can be defined, which are on the first place for the government in terms of MOOC development and attraction of the more vast masses to them:

- Moscow State University;
- Higher School of Economics;
- Saint Petersburg State University;
- Moscow Institute of Physics and Technology;
- Russian Academy of the National Economy and Civil Service at the RF President.

Some of the above-stated universities already collaborate with the international project such as Coursera, however there is need for Russian projects development. For now, there are MOOC that are organized directly from the Russian universities and organization in fields of education and culture resources:

- Universarium (www.universarium.org) [11]
- Eduson (www.eduson.tv) [12]
- Uniweb (www.uniweb.ru) [13]
- Univer.tv [14]
- Intuit.ru (National Open University INTUIT) [15] and others.

4. Conclusion

To sum up, it should be noted that the process of e-learning implementation in Russian higher education system proceeds slowly, facing particular difficulties: electronic courses quality problem (who and how can access them), legal issues, related to the intellectual property protection, financial ones, concerning the electronic courses preparation costs and their renovation, recruitment problems, related to the training of teachers who are able and want to develop and instantly renew these courses. The low confidence level to the distant form of education in academic circles should also be referred to the factors restraining online education development. The historically developed opinion that the traditional education form is the most effective one still exists in Russian minds. The mass realization process of the all advantages of such education form will take a lot of time. To the other reasons of Russian universities lag in the new education form process can also be referred: the lack of necessary normative base in the distant education sphere; the absence of specialists in electronic courses development sphere; the shortcoming of financial resources needed for electronic educational content development; the absence of teachers readiness to use distant technologies in their work; issues of copyright law and related to it teachers reluctance to publish their recourses in open access.

However, the obvious thing that electronic education is not a temporary passion and we need to care about complex solution of problems mentioned above already today, to prepare teachers and students to use Internet global network and the electronic education facilities as a whole.

References

[1] [SibSAU Distance Learning Server]. (In Russ.). Available at: http://dl.sibsau.ru (accessed 05.10.2015)

[2] Satunina A.E. [E-learning: advantages and disadvantages]. Sovremennye problem nauki I obrazovaniya. 2006, no. 1, p. 89-90. (In Russ.)
[3] [Coursera]. (In Russ.). Available at: http://www.coursera.org. (accessed 04.03.2016)
[4] [Futurelearn]. (In Eng.). Available at: http://www.futurelearn.com (accessed 04.03.2016)
[5] [Cambridge English Teacher]. (In Eng.). Available at: http:// www.cambridgeenglishteacher.org (accessed 04.03.2016)
[6] [EdX]. (In Eng.). Available at: http:// www.edx.org (accessed 04.03.2016)
[7] [Udacity]. (In Eng.). Available at: http:// www.udacity.com (accessed 04.03.2016)
[8] [Khan Academy]. (In Eng.). Available at: http://www.khanacademy.org (accessed 04.03.2016)
[9] [Moscow School of Management SKOLKOVO]. (In Russ.). Available at: http://www.skolkovo.ru (accessed 04.03.2016)
[10] [Education in Russia for Foreigners]. (In Russ.). Available at: http://www.russia.edu.ru (accessed 04.03.2016)
[11] [Universarium]. (In Russ.). Available at: http://www.universarium.org (accessed 04.03.2016)
[12] [Eduson]. (In Russ.). Available at: http://www.eduson.tv (accessed 04.03.2016)
[13] [Uniweb]. (In Russ.). Available at: http://www.uniweb.ru (accessed 04.03.2016)
[14] [Univer.tv]. (In Russ.). Available at: http://www.univer.tv (accessed 04.03.2016)
[15] [Intuit.ru]. (In Russ.). Available at: http://www.intuit.ru (accessed 04.03.2016)

Individual characteristics of language learners in lesson planning for specific purposes

N V Ivleva
Business Foreign Languages Department
Reshetnev Siberian State Aerospace University,
31, Krasnoyarsky Rabochy Av., Krasnoyarsk, 660037, Russian Federation

E-mail: natalie_ivleva@mail.ru

Abstract. This paper provides a summary of internationally recognized methods and approaches in teaching foreign languages with the focus on teaching foreign languages for specific purposes. It also covers the idea of mastering a foreign language through knowing speech units which are necessary for individual language learners in the professional area with due regard to criteria provided by the author. The criteria were developed on the basis of the theory on adaptation of complex systems and also observation in the teaching process.

Key words. Individual characteristics, language learners, adaptation, speech units.

1. Introduction

Learning a foreign language is a priority in the professional advancement and development of a specialist in any area. It is stipulated by the processes of globalization, which oblige the society to provide information about technologies, designs and other activities implemented through foreign languages. Any professional area, having access to the international community, fulfills all the processes, or partially, in the English language. Consequently, to share the experience with foreign colleagues, undertake an internship program abroad, get opportunities to carry out activities with the focus on foreign technologies, implement those technologies and other innovations on the territory of a home country it is crucial to master a foreign language, especially English, as it is considered to be the international one. Due to abovementioned facts there are reasons for English popularization in the professional community and increasing the number of people, learning the language.

In response to demand there is supply of educational services specializing in giving English language courses. Such services are provided both by private and state educational institutions within fee-based educational services. Competition on this market is rather big in any major population center. However, not many of such institutions are able to define the effective educational path for a specialist applied to them. Currently, a great number of methodologists emphasize teaching methods and approaches. They reckon that in case the right method is chosen, it will be faster for a learner to master a foreign language. The other group of methodologists states that except for methods and approaches it is vital to pay attention to individual characteristics of a specialist who applied for getting skills in a foreign language.

2. Methods and approaches in teaching foreign languages

According to Jim Scrivener, ELT author and teacher trainer, there are 12 methods and approaches in teaching foreign languages recognized worldwide:

Grammar-Translation Approach (GTA). Much traditional language teaching in schools worldwide used to be done in this way, and it is still the predominant classroom method in some cultures. The teacher rarely uses the target language. Students spend a lot of time reading texts, translating them, doing exercises and tests, writing essays. There is relatively little focus on speaking and listening skills.

Audio-Lingual Method (ALM). Although based on largely discredited theory, the techniques and activities continue to have a strong influence over many classrooms. It aims to form good habits through students listening to model dialogues with repetition and drilling but with little or no teacher explanation.

Communicative Approach (CA). This is perhaps the method or approach that most contemporary teachers would subscribe to, despite that fact that it is widely misunderstood and misapplied. CA is based on beliefs that learners will learn best if they participate in meaningful communication. It may help if we distinguish between a stronger and a weaker vision CA. As a strong aspect students learn by communicating, but they also learn through a wide variety of teaching, exercises, activities and study, with a bias towards speaking and listening work.

Total Physical Response (TPR). A method devised by Dr. J. Asher, mainly useful with beginner and lower-level students. Learners listen to instructions from the teacher, understand and do things in response, without being required to speak until they are ready.

Community Language Learning (CLL). A method based around use of the learners' first language and with teacher help in mediating. It aims to lower anxiety and allow students to communicate in a more genuine way than is typically possible in classrooms.

Natural Approach (NA). Devised by Stephen Krashen, this is a collection of methods and techniques from many sources, all intended to provide the learner with natural comprehensible language so that the learner can pick up language in way similar to a child learning their first language.

Task-Based Learning (TBL). A variant of TBL which bases work cycles around the preparation for, doing of, and reflective analysis of tasks that reflect eral life needs and skills.

Silent Way (SW). Devised by Caleb Gattegno, this method requires the learner to take active ownership of their language learning and to pay great attention to what they say. Distinctive features include the relevant restraint of the teacher and the use of specially designed wall charts. The use of Cuisenaire rods in mainstream ELT arose from this method.

Person-Centered Approach (PCA). Any approach that places learners and their needs at the heart of what is done. Syllabus and working methods will not be decided be the teacher in advance of the course, but agreed between learner and teacher.

Lexical Approach (LO). Proposed by Michael Lewis and Jimmie Hill. On the back of new discoveries about how language is really used, especially the importance of lexical chunks in communication, proponents suggest that traditional present-then-practise methods are of little use and propose a methodology based around exposure and experiment.

Dogme. Scott Thornbury's proposed back-to-basics approach. Teachers aim to strip their craft of unnecessary technology, materials and aids and get back to the fundamental relationship and interaction of teacher and student in class.

Content and Language Integrated Learning (CLIL). This approach refers to teaching subjects such as science, history and geography to students through a foreign language. This can be by the English teacher using cross-curricular content or the subject teacher using English as the language of instruction. Both methods result in the simultaneous learning of content and English [1].

However, there is one more approach - the Eclectic Approach - which allows teachers to use techniques and activities from a range of language teaching methods and approaches. The teacher decides which method or approach to use depending on the aims of the lesson and the learners in the group. For example, the class starts with an inductive activity with learners identifying the different uses of synonyms of movement using a reading text. They then practise this using Total Physical Response. In another class the input is recycled through Task-Based Learning, with learners producing the instructions for an exercise manual.

The eclectic approach does not mean random, unmethodical usage of various teaching methods and approaches. It is better to name this approach the integrated one as this term has positive connotation. Using the integrated approach can be justified if elements from different methods and approaches are formed into one logical complex system and encourage more effective language learning. For example, it is a good way to combine the Communicative and Task-Based Learning methods, Lexical Approach and Content and Language Integrated Learning method, etc.

The Eclectic Approach is widely used in educational programs which are integrated in fact as they contain structures and lexical units typical for the grammar-translation and lexical approaches, speech functions, used in the communicative approach, a list of skills, abilities and other components which are usually included into programs. Such programs are more effective as they cover the whole study process in different countries.

Taking all descriptions of abovementioned approaches and methods into account while lesson planning it is important to choose those which are going to work with a certain target audience and meet their needs finally. Furthermore, such an approach will be productive only if all the individual characteristics of a specialist are noted [2].

3. Speech units and their introduction into study process

According to Leonard A. Rastrigin, doctor of technical sciences, professor, contributed a lot into development of such scientific areas as cybernetics and informatics, the study process must be considered as a complex system. A standard task for a teacher is to arrange a study process as well as it could be to memorize a certain portion of information. In our case by a certain portion of information we mean speech units. Speech units are a combination of language units in the speech chain. The algorithm of teaching a foreign language presents the right choice of speech units to be realized and learnt as a portion of information. The efficiency of such teaching can be assessed with the help of a level test both in written and oral forms. It can also be visible in the process of using the language in practice [3].

Relying on the theory offered by Leonard A. Rastrigin and observations in the frame of teaching activity it is possible to state that teaching efficiency directly depends on individual characteristics of a language learner, which means – speech units he/she is in need of for communication in the professional area – and competently chosen teaching methods or approaches. As soon as a teacher finds out necessary speech units and applies the best teaching methods and approaches efficiency will level up and be noticeable in practice.

In connection with abovementioned criteria it is offered to use a list of criteria for defining speech units to be introduced into a study process which are: the professional area of a learner, the form of communication in a foreign language (business writing, oral communication, translation or interpretation), the study period, goals a learner sets and also a memorization technique (associations, repetition, writing, etc.).

Depending on a professional area of a learner a teacher is able to choose appropriate topics within which certain speech units are introduced and learnt. It is possible to arrange a study process efficiently in case of motivation in a topic discussed at the lesson and for future stimulation in mastering a foreign language. A form of communication also plays a crucial role upon designing and planning lessons. A lot of language learners need to communicate with their partners and clients in a written form (by email) more than orally. Due to this fact it is important to draw learners' attention to functional speech units and also accuracy and cohesion in using them. If a learner communicates orally in most cases then it is better to focus on vocabulary. In such situations it is also recommended to brush up grammar structures and introduce the new ones only in case of needs which must be met in practice. To introduce advanced structures as new speech units is not logical as it can lead to wasting time and extending terms of study that can be critical for the majority of learners. It is also vital to realize the final goals a language learner sets in terms of the study process. It is common for many learners that they would like to master a language just to know it. Such a "goal" rarely leads to proper results as there is no proper goal and, consequently, no proper results. In similar situations a teacher should assist in setting goals knowing abovementioned parameters. There is a wide variety of goals such as orientation to international language exams, travelling abroad or motivation for communication in a foreign community. The last but not the least criterion is to understand how a learner is able to memorize information in his life and which techniques he uses in his studies. There are a reasonable number of people who use associations to memorize information. The idea of this technique is to connect or correlate a new speech unit to a well-known object, activity or definition to create a logical chain of speech units which helps effective memorizing. Another category of people memorizes information only through making notes. This technique also works in language learning.

4. Conclusion

In general, to understand what a teacher should pay attention to it is important to analyze a future language learner from the point of described criteria running an interview with them or asking them to participate in a survey, having chosen the best teaching methods or approaches. During the study process a teacher should continue observing and adapting the educational path regarding language mastering.

References

[1] Jim Scrivener. Learning Teaching. The Essential Guide to English Language Teaching, 3rd edition, Macmillan, 2011. – p. 418.

[2] Kashina E.G. Traditsii I innovatsii v metodike prepodavaniya inostrannogo yazyka / uchebnoe posobie dlya studentov filologicheskikh fakultetov universitetov, "Univers-grupp", 2006.
[3] Rastrigin L.A. Adaptatsiya slozhnykh system, RIGA "Znanie", 1981. – p. 55.

Spatial suppression of interference in hybrid reflector antennas *

I N Kartsan[1] P V Zelenkov[2] V N Tyapkin[3] D D Dmitriev[4] A E Goncharov[5]
[1] Associate professor, Reshetnev Siberian State Aerospace University, Krasnoyarsk, Russia
[2] Science and Research Department Director, Reshetnev Siberian State Aerospace University, Krasnoyarsk, Russia
[3] Associate professor, Siberian Federal University, Krasnoyarsk, Russia
[4] Associate professor, Siberian Federal University, Krasnoyarsk, Russia
[5] Associate professor, Reshetnev Siberian State Aerospace University, Krasnoyarsk, Russia

E-mail: kartsan2003@mail.ru

Abstract. The article describes a 7-beam hybrid reflector antenna with a 19-element feed element which forms a radiation pattern in the form of a cluster. During the research the synthesis of the amplitude-phase distribution of the antenna feed element has been conducted. A radiation pattern for three situations of interference: along the first side lobe, along the ramp of the main lobe, along the main lobe.

Keywords: hybrid reflector antenna, special filtering, interference, adaptive algorithms, antenna radiation pattern

1. Introduction

Currently, more attention is paid to the issue of satellite communication systems in Russia. The primary reason for this is the necessity to prove reliable communication to distant areas of the country: especially for the underpopulated areas of Siberia, Yakutia, the Far East, and the Russian Arctic. Considering the unequally populated territory of Russia (the majority of the population resides in the European part of the country or in southern areas of Siberia) and the large number of time zones, it is necessary to include a flexible control system for the communications channels. This flexibility is basically a traffic redistribution system, which functions depending on the load. Another factor for developing such a system is the high concentration of radiating radio-electronic devices in the industrial parts of the country: this often causes problems with electromagnetic compatibility. There are many cases when industrial interference caused malfunction in satellite communication systems. This makes it urgent to include interference protection for communications channels in all new satellite communications systems.

Among the more perspective methods for solving this problem is the utilization of multibeam adaptive phased antenna arrays for an onboard antenna. By retargeting the beams, these antennas enable effective flexible control of the traffic. Also they can provide special filtering by controlling the radiation pattern: forming maximums targeted at the source of the signal and minimums at the source of interference. However, multibeam adaptive phased antenna arrays have a number of disadvantages. They are quite expensive – this is due to the

* This research has been financed by the Ministry of Education and Science of the Russian Federation (contract No. 14.577.21.0155, unique project identifier RFMEFI57714X0155).

complex production process for these devices, they have a large mass and are take up much space, the amplification coefficient is low. Hybrid reflector antennas with adaptive phased antenna arrays for a feed element are possessed of these disadvantages to a lesser extent. Utilizing hybrid reflector antennas in spacecraft in comparison with adaptive phased antenna arrays is more perspective – especially in terms of lower costs and reduction of weight – due to higher requirements to the amplification coefficient of the antenna system (exceeding 35 – 40 dB) [1]. In this case, to construct an adaptive phased antenna array a large number of elements from 90 to 255 is necessary for a radiation pattern of 2°×2° и 1°×1°. This results in significant in the increase of productions costs (up to 4 times) and mass (80 to 100 kg). The construction of the array also becomes overengineered. Adaptive hybrid reflector antennas have lesser scanning angles, however for geostationary spacecraft this factor is not a significant disadvantage and is substituted by the simpler design and lower mass (30 – 40 kg).

One of the main problems for building adaptive multibeam hybrid reflector antennas is the design of methods and algorithms to spatial adaptiveness to interferences. In [2–5] an algorithm for synthesizing amplitude-phase distribution of hybrid reflector antennas has been designed. This algorithm considers the features of both adaptive phased antenna arrays and reflector antennas.

2. Modelling the synthesis method for amplitude-phase distribution of a radiating array of a hybrid reflector antenna during interference

Let's consider an adaptive multibeam antenna based on a hybrid reflector antenna. This antenna consists of a mirrored reflector and feed element in the form of a 19-element antenna array that generates 7 beams of the radiation pattern on the basis of a cluster. Each cluster, consisting out of 7 feed elements is powered cophasally: half of the radiated power is transmitted on to the central feed element and half is transmitted to the peripheral feed elements [1, 2].

Let's perform a modelling of the amplitude-phase distribution of a radiating array of a hybrid reflector antenna during interference. The radiation pattern of the adaptive hybrid reflector antenna is calculated using GRASP 9.3.03 and SATSOFT software. This software is certificated and verified for performing calculations of the radiation pattern for antennas.

The results of calculating the radiation pattern for a K_u band hybrid reflector antenna is shown in Fig. 1. The thin line shows the required service zone 1°×1°.

To provide independent functioning of each beam of the adaptive hybrid reflector antenna, a frequency division is arranged between the beams (Fig. 2). Dividing the beams with seven frequencies is the most convenient: it enables to get an interbeam discrimination between 25 and 39 dB and fully eliminate the influence of beams on one another both in the mode when interference is absent and the suppression mode.

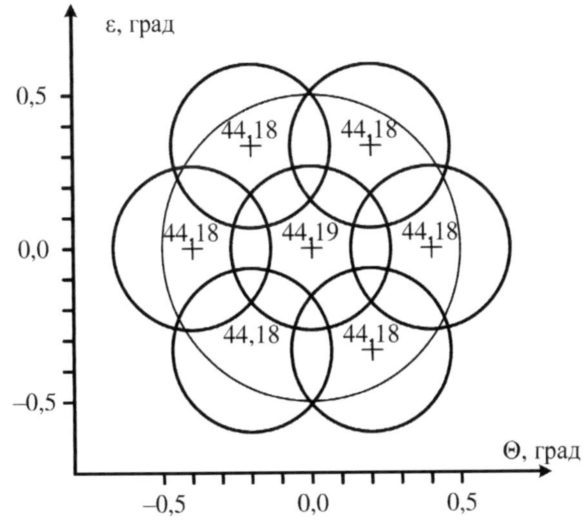

Figure1. Covering the service area with asset of beams. Performance amplitude is 43.5 GHz; amplification coefficient for maximum radiation pattern of focal beam is 44.19 dB; amplification coefficient for maximum radiation pattern of peripheral beam is 44.16 dB.

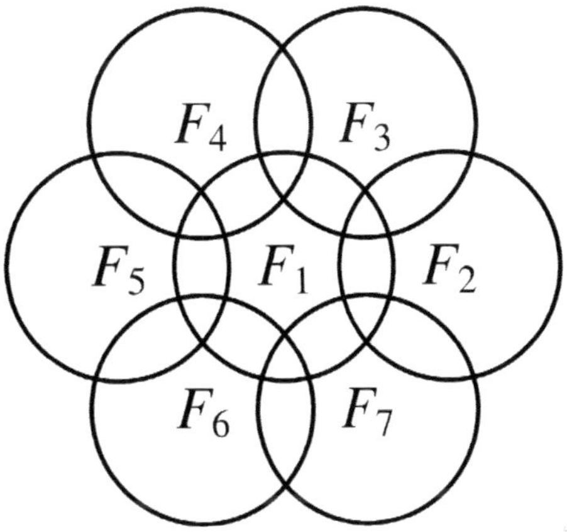

Figure 2. Frequency division between adaptive hybrid reflector antenna beams.

Let's perform calculations for interference suppression based on the proposed radiation pattern synthesis method in [4, 5]. The calculation is performed for the central beam of the adaptive hybrid reflector antenna. Since each beam of the antenna functions independently, similar results are achieved for the rest of the beams.

Figs. 3 – 5 demonstrate the synthesis results of amplitude-phase distribution of a radiating array of a hybrid reflector antenna. Its calculated radiation patterns are shown in Table 1. Synthesis of the radiation pattern in performed for three cases of directing interference effect:

- interference in direction of the first side lobe of the antenna;
- interference in direction of the ramp of the main lobe of the antenna;
- interference in direction of the main lobe of the antenna.

To synthesize the amplitude-phase distribution cluster, which consists out of 7 elements approximately 400 iterations of the algorithm were required, the number of points for the amplitude-phase distribution synthesis is 16. The position of the interference was determined by the program.

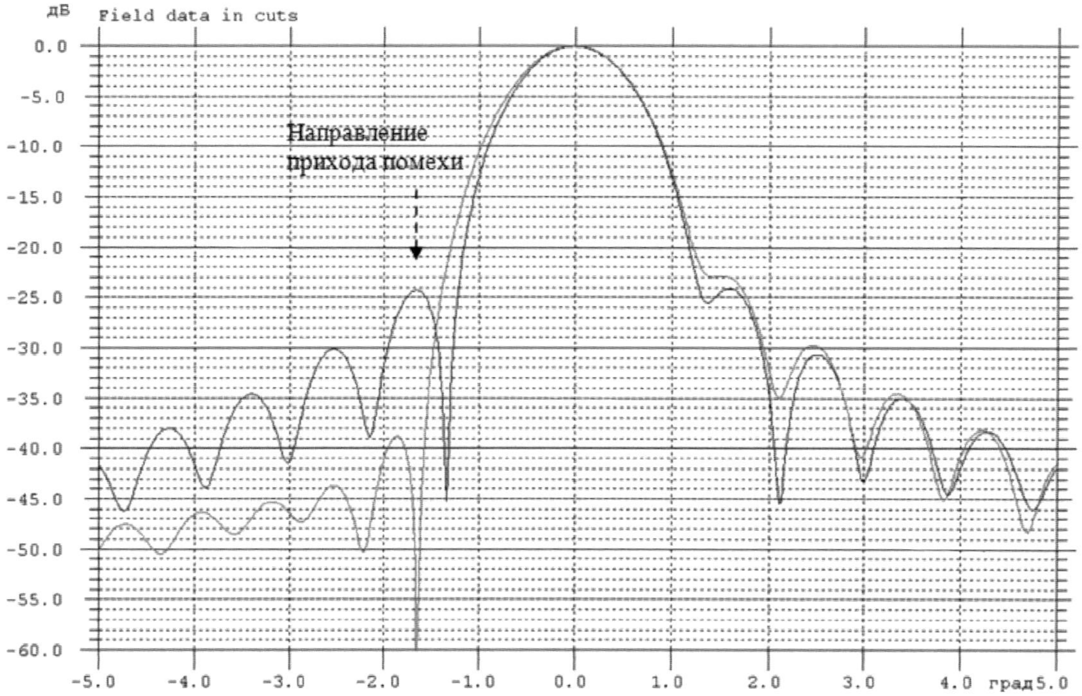

Figure 3. Results of radiation pattern synthesis for interference suppression in direction of first side lobe.

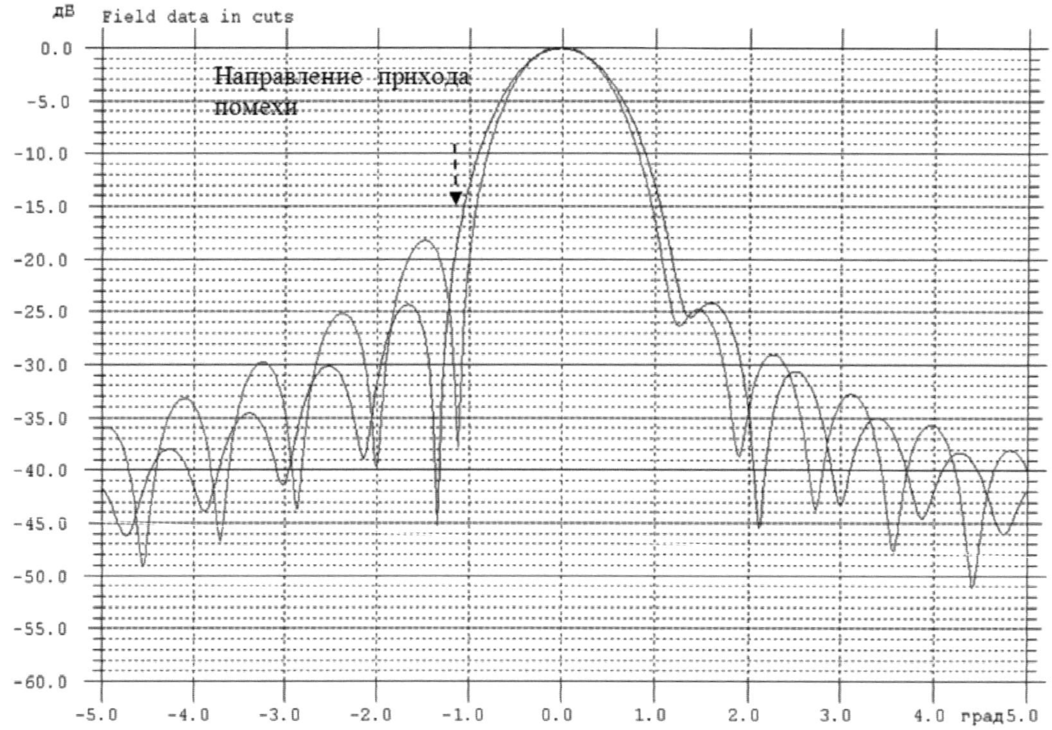

Figure 4. Results of radiation pattern synthesis for interference suppression in direction of ramp of main lobe of radiation pattern.

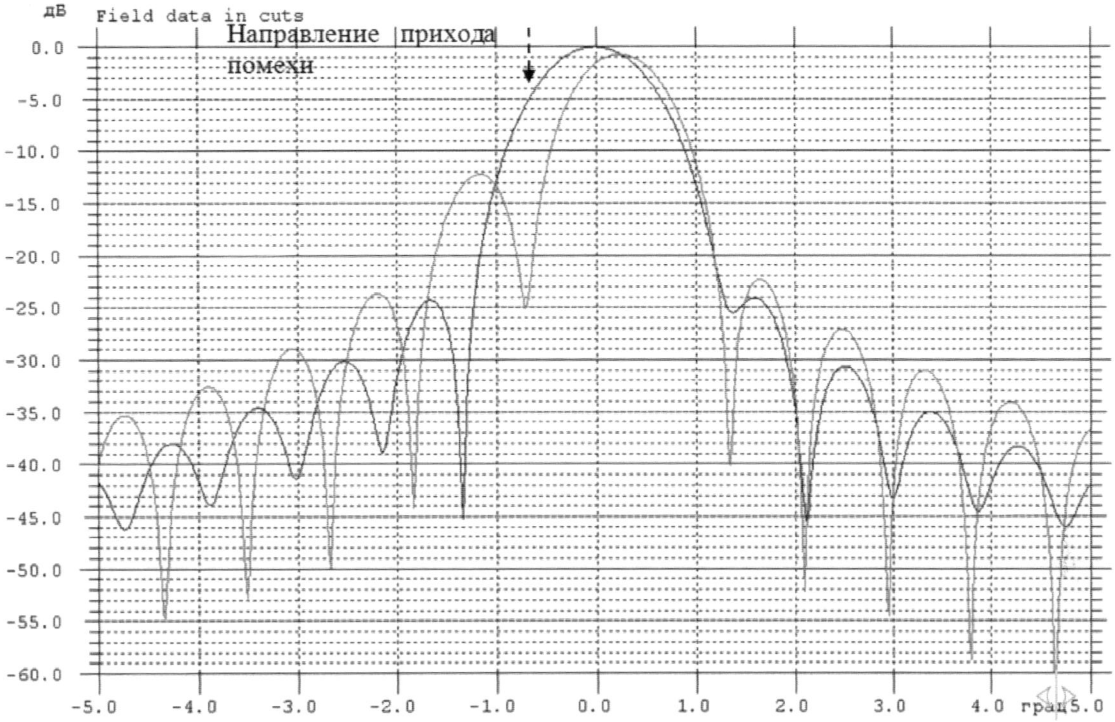

Figure 5. Results of radiation pattern synthesis for interference suppression in direction of main lobe of radiation pattern.

Table 1. Results of amplitude-phase distribution synthesis for radiation pattern.

Number of feed element	Direction of interference effect					
	First side lobe		Ramp of main lobe		Main lobe	
	Amplitude dB	Phase, grad	Amplitude dB	Phase, grad	Amplitude dB	Phase, grad
1	–5.33	–2.84	–2.54	–177.85	–6.39	–171.04
2	–7.72	1.08	–34.6	–171.74	–13.26	12.01
3	–10.88	0.32	–7.3	178.22	–5.19	145.05
4	–8.65	0.65	–11.86	–179.73	–8.09	161.23
5	–10.23	0.05	–14.58	–165.55	–13.36	–150.72
6	–8.49	1.01	–10.19	–164.82	–9.36	–125.02
7	–10.79	1.14	–12.19	–164.37	–9.86	–75.31

The calculation results for the radiation pattern of an adaptive hybrid reflector antenna demonstrate that during interference suppression in direction of the ramp of the main lobe the level of interference decreases to minus 25 – 35 dB while the signal level remains unchanged in the rest of the service area. During interference suppression in direction of the main lobe of the radiation pattern, a reduction in the amplification coefficient to 1 dB is observed; the interference suppression level is recorded at minus 25 dB – a satisfactory result for satellite communications systems.

For the given scheme of adaptive hybrid reflector antenna arrangement it is possible to utilize a functioning mode in which each antenna beam will suppress interference independently from the other beams. Since each antenna beam is formed by seven feed elements the amplitude-phase distribution synthesis will enable interference to be suppressed in a similar to the 7-element cluster way. This method is optimal considering the system's effectiveness in whole and signal level maintenance for the remaining service area.

Let's suppose that we are sure of the interference positions in advance. Since the beams function independently from one another and are divided by frequency, an amplitude-phase distribution synthesis for the 7-element cluster is performed for each beam to suppress interference.

Figs. 6 – 11 demonstrate the modelling results for suppressing interference for each beam of the adaptive hybrid reflector antenna, the neighboring beams without interference adaptiveness.

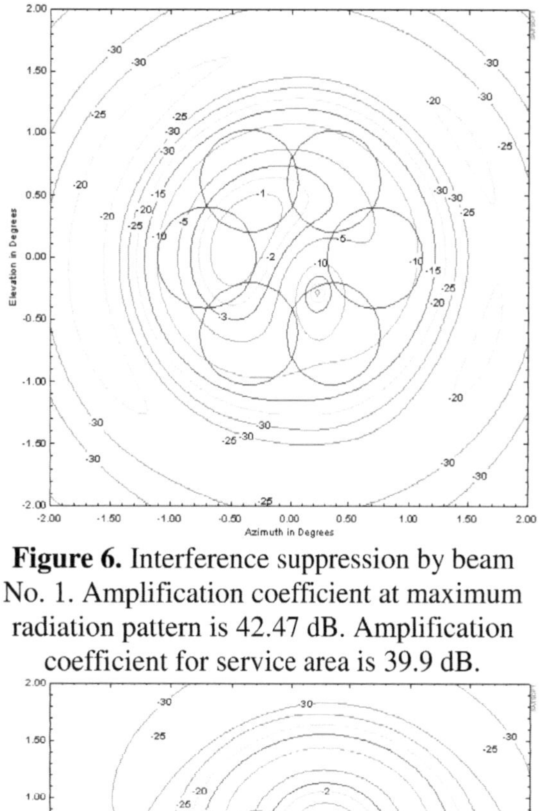

Figure 6. Interference suppression by beam No. 1. Amplification coefficient at maximum radiation pattern is 42.47 dB. Amplification coefficient for service area is 39.9 dB.

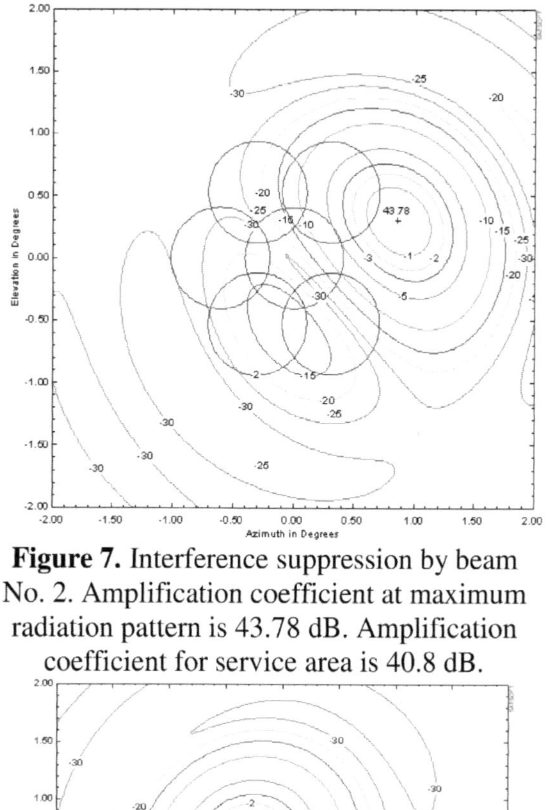

Figure 7. Interference suppression by beam No. 2. Amplification coefficient at maximum radiation pattern is 43.78 dB. Amplification coefficient for service area is 40.8 dB.

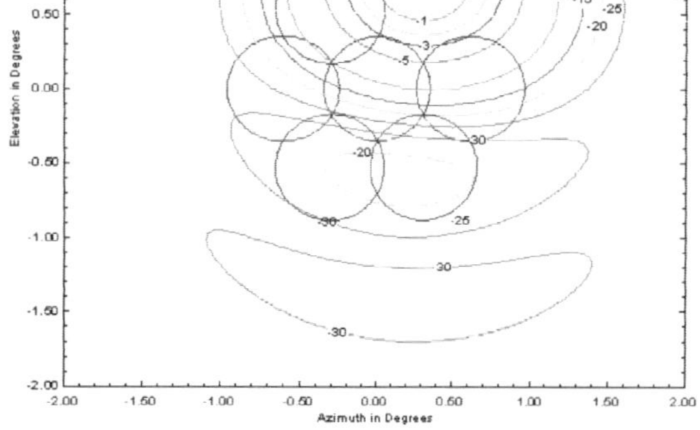

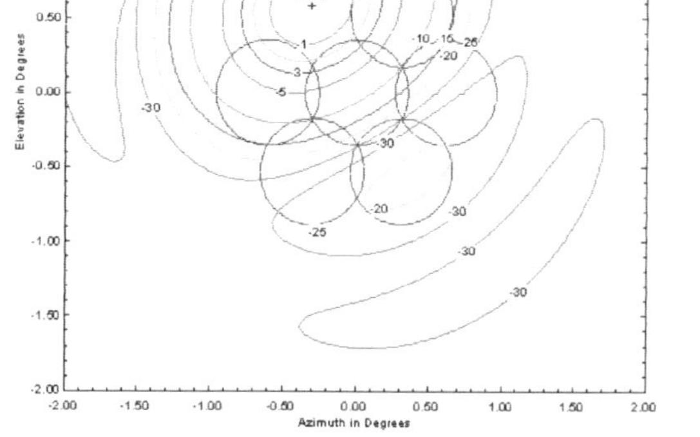

Figure 8. Interference suppression by beam No. 3. Amplification coefficient at maximum radiation pattern is 44.55 dB. Amplification coefficient for service area is 41.5 dB.

Figure 9. Interference suppression by beam No. 4. Amplification coefficient at maximum radiation pattern is 44.6 dB. Amplification coefficient for service area is 41.6 dB.

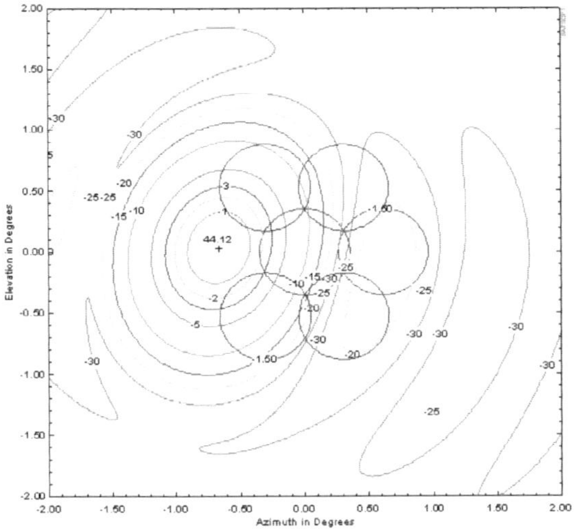

Figure 10. Interference suppression by beam No. 5. Amplification coefficient at maximum radiation pattern is 42.12 dB. Amplification coefficient for service area is 41.1 dB.

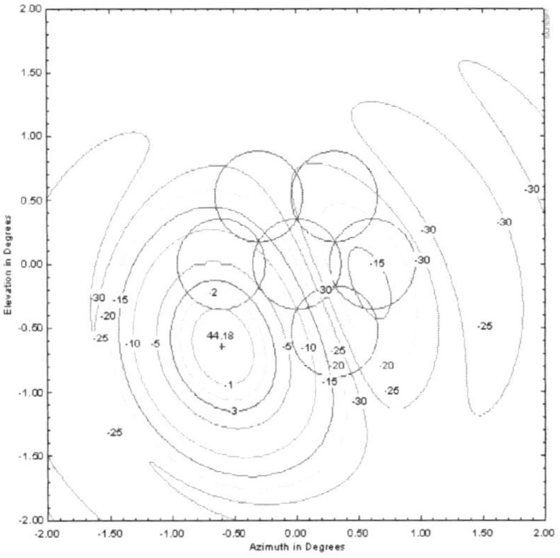

Figure 11. Interference suppression by beam No. 6. Amplification coefficient at maximum radiation pattern is 44.18 dB. Amplification coefficient for service area is 40.2 dB.

3. Conclusion

A radiation pattern for suppressing interference to a level of minus 40 dB in relation to the radiation pattern maximum has been constructed for each beam of an adaptive hybrid reflector antenna, built on the cluster scheme, using the results of the amplitude-phase distribution synthesis. The signal level for the remaining service area was 39.9 dB for the worst beam.

Due to the frequency division of the beams, the adaptiveness of the given seven beams will not influence the neighboring beams since the radiation pattern has high amplitude for decreasing signal level. This provides required a discrimination between beams with the same frequency.

References

[1] Tyapkin V N, Pershin A S, Dmitriev D D, Moshkina T G 2012 Adaptive multibeam antennas *Vestnik of Siberian State Aerospace University named after academician M. F. Reshetnev* Issue 2 (42) P 106–109

[2] Tyapkin V N, Dmitriev D D, Perschin A S 2013 The adaptive algorithm multibeam antenna, built on hybrid-reflector antenna *Journal of Siberian Federal University. Engineering & Technologies, No. 7 Vo. 6 p 835–844*

[3] Tyapkin V N, Dmitriev D D, Garin E N, Sokolovskii A V 2013 Controlling amplitude-phase distribution with an adaptive phase antenna array *Promyshlennye ASU i kontrollery* No. 4 P 66-71

[4] Tyapkin V N, Kartsan I N, Dmitriev D D, Goncharov A E 2015 Spatial filtering algorithms in adaptive multi-beam hybrid reflector *antennas International Siberian Conference on Control and Communications (SIBCON) – Proceedings*
[5] Serenkov V I, Kartsan I N, Dmitriev D D 2015 Synthesis methods for amplitude-phase distribution for hybrid reflector antennas *Vestnik of Siberian State Aerospace University* Vol 16 No. 3 P 664-669

IOP Conf. Series: Materials Science and Engineering **122** (2016) 012011 doi:10.1088/1757-899X/122/1/012011

Greedy heuristic algorithm for solving series of eee components classification problems*

L A Kazakovtsev[1,2], A N Antamoshkin[1], V V Fedosov[3]
[1]Siberian State Aerospace University named after Academician M.F.Reshetnev
31 "KrasnoyarskiyRabochiy" prospect, Krasnoyarsk, 660037, Russia.
[2]Siberian Federal University,
79, Svobodny Prospect, Krasnoyarsk, Russia
[3]"TTC – NPO PM" JSC
20, Molodezhnaya ul., Zheleznogorsk, Krasnoyarskiy kray, 662970, Russia

E-mail: levk@bk.ru

Abstract. Algorithms based on using the agglomerative greedy heuristics demonstrate precise and stable results for clustering problems based on k-means and p-median models. Such algorithms are successfully implemented in the processes of production of specialized EEE components for using in space systems which include testing each EEE device and detection of homogeneous production batches of the EEE components based on results of the tests using p-median models. In this paper, authors propose a new version of the genetic algorithm with the greedy agglomerative heuristic which allows solving series of problems. Such algorithm is useful for solving the k-means and p-median clustering problems when the number of clusters is unknown. Computational experiments on real data show that the preciseness of the result decreases insignificantly in comparison with the initial genetic algorithm for solving a single problem.

Introduction

Supplying the electronic units of the complex technical systems with the highest quality EEE components increases the whole system reliability. Moreover, for reaching the highest reliability of an electronic unit, the EEE components of the same type must have equal characteristics which assure their coherent operation. The highest homogeneity of the characteristics is reached if the EEE components are produced as a single production batch from a single batch of the raw materials [1]. The critically important units are integrated from EEE components manufactured as a special production lots with special quality requirements [2, 3].

The characteristics of each EEE device in the lot are checked using destructive and nondestructive tests [2, 4]. Resulting data of such tests are used for analyzing the lot homogeneity [4]. For splitting the EEE components into several assumed production batches, the k-means method is used [5, 6, 7].

* Results were obtained in the framework of the state task № 346 of the Ministry of Education and Science of the Russian Federation

Manufacturers of the EEE components in the United States of America and European Union produce components of special quality classes, Military and Space [8, 9]. Manufacturers in the Russian Federation do not form a special class of components for use in space systems [2, 3].

The k-means problem can be classified as a continuous problem of the location theory [10, 11, 12]. The aim is to find k points (centers, centroids) in a d-dimensional space such that the total squared distance from each of the data vectors (known points, measurement result vectors) to the nearest of k chosen centers reaches its minimum:

$$\arg\min F(X_1,\dots,X_k) = \sum_{i=1}^{N} \min_{j\in\{1,k\}} \|X_j - A_i\|^2.$$

The aim of a continuous p-median problem (also known as k-median problem) [11] is finding k points (called centers or medians) such that the sum of weighted distances from N known points (data vectors) to the nearest of k centers reach its minimum.

$$\arg\min F(X_1,\dots,X_k) = \sum_{i=1}^{N} w_i \min_{j\in\{1,k\}} L(X_j, A_i).$$

Here, wi are the weight coefficients, $L()$ is some distance function between two given poins. Usually, $L()$ is a metric in the d-dimensional space.

Continuous location problems with Euclidean, Manhattan (rectilinear), Tschebychev metrics a well investigated (all such metrics are special cases of metric based on Minkovski l_p-norms [13]), many authors propose algorithms for solving p-median problems with such metrics and their special case, 1-median problem (also called the Weber problem) which is used for solving the p-median problem. In particular, well known Weiszfeld procedure [44] was generalized for metrics based on Minkovski norms.

Traditionally, the problem with the Euclidean metric $L(X_j,A_i)=\sqrt{\sum_{k=1}^{d}(x_{j,k}-a_{i,k})^2}$ is called p-median problem. Here, $X_j=(x_{j,1},\dots,x_{j,k})\ \forall j=\overline{1,p}$, $A_i=(a_{i,1},\dots,a_{i,k})\ \forall i=\overline{1,N}$. In case of the squared Euclidean metric $L(X_j,A_i)=\sum_{k=1}^{d}(x_{j,k}-a_{i,k})^2$ and weight coefficients $w_i=1 \forall i=\overline{1,N}$, we have the k-means problem.

Known methods

The k-means method uses the ALA procedure (Alternating Location-Allocation) which includes two simple steps:

<u>Algorithm 1.</u> ALA procedure.

Required: data vectors $A_1 \dots A_N$, k initial cluster centers $X_1 \dots X_k$.

1. For each center X_i, determine its cluster C_i as a subset of the data vectors for which this center X_i is the closest one.
2. For each cluster C_i, recalculate its center X_i (i.e., solve the Weber problem).
3. Repeat Step 1 unless Steps 1, 2 made no change in any cluster.

Except special cases, the k-means and p-median problems are NP-hard and require global search [15].

The result of the ALA procedure depends on the choice of initial cluster centers. Known k-means++ algorithm [16] has an advantage in comparison with the random choice of the initial centers with uniform distribution. However, this improvement preciseness is insufficient for many practically important problems which need more precise results. For such cases, researchers propose various recombination techniques for initial center sets [10].

The ALA procedure can be optimized with use of many techniques. For example, sampling procedures [17] solve the k-means problem for the randomly selected subset of the data vectors and use the achieved result as an initial set of centers for solving the original problem.

The dependence of the results of the ALA procedure on the initial centers seeding is a serious problem for the reproducibility of the classification algorithm results: depending on the initial centers seeding, different algorithm starts classify the same data vectors as elements of various clusters. For the EEE component production batches classification problem, this means that various EEE component belong to the same or different production batches depending on the initial seeding. Thus, an algorithm for solving k-means problem which returns a stable result is preferred.

Rather precise but extremely slow, the Information Bottleneck Clustering method (IBC) is a deterministic method for solving the cluster analysis and classification problems able to achieve perfect results in many cases.[18]. This algorithm starts from considering each data vector as a separate cluster. Then, clusters are removed one-by-one until the desired clusters quantity remains. Each time, the algorithm eliminates such cluster that its elimination gives the smallest increment of the objective function value. For the k-means and p-median problems, this algorithm eliminates the cluster center which gives the smallest total distance from data vectors to the closest remaining centers.

The genetic algorithms (GAs) with greedy agglomerative heuristic initially designed for the discrete k-median problem on a network [19] are compromise variants by preciseness, stability of the results and time consumption. In [20, 12], author propose an approach for adaptation of these algorithms for the continuous location problems:

Algorithm 2. GA with greedy heuristic and floating point alphabet.

Required: Set $V = (A_1,\dots,A_N) \in \mathbb{R}^d$, quantity of clusters p, GA population size N_p.

1: Create N_p sets of coordinates $\chi_1,\dots,\chi_{N_p}$: $\chi_i \subset \mathbb{R}^d$, $|\chi_k| = p \forall k = \overline{1,N_p}$ whica are results of the ALA procedure. Thus, each χ_i is a local minimum of the solved problem. Store the values of the objective functions to an array of variables $\mathcal{F}_1,\dots,\mathcal{F}_{N_p}$.

2: If the stop conditions are reached then go to Step 8.

3: Randomly choose two "parent" sets χ_{k_1} and χ_{k_2}, $k_1, k_2 \in \{\overline{1,N_p}\}$, $k_1 \neq k_2$. Running algorithm 3, obtain the "child" sets of coordinates χ_c which are local minimums of the objective function. Store the value of the objective function $\mathcal{F}_c$.

4: If $\exists k \in \{\overline{1,N_p}\}$: $\chi_k = \chi_c$ then go to Step 2.

5: Choose an index $k_{worst} = \mathrm{argmax}_{k=\overline{1,N_p}} \mathcal{F}_k$. If $\mathcal{F}_{wotst} < \mathcal{F}_c$ then go to 2.

6: Randomly choose two indexes k_1 and k_2, $k_1 \neq k_2$; array $k_{worst} = \mathrm{argmax}_{k\in\{k_1,k_2\}} \mathcal{F}_k$.

7: Swap the values of $\chi_{k_{worst}}$ and χ_c, store $\mathcal{F}_{k_{worst}} = \mathcal{F}_c$ and go to Step 2.

8: STOP. Result is set χ_k^*, $k^* = \mathrm{argmin}_{k=\overline{1,N_p}} \mathcal{F}_k$.

Modification of the greedy agglomerative heuristic procedure for this GA is as follows.

Algorithm 3. Greedy crossingover heuristic for Algorithm 2.

Required: Set $V = (A_1, \dots, A_N) \in \mathbb{R}^d$, quantity of clusters p, two "parent" sets of centers χ_{k_1} and χ_{k_2}, values σ_e and L_{min}.

1: Join sets $\chi_c = \chi_{k_1} \cup \chi_{k_2}$. Run the ALA procedure for $|\chi_c|$ clusters starting from the solution χ_c. Store its result in χ_c.

2: If $|\chi_c| = p$, then run the ALA procedure with the initial solution χ_c, then STOP and return result χ_c.

2.1: Calculate the distances from each data vector to the nearest element of χ_c.

$$d_i = \min_{X \in \chi_c} L(X, A_i) \forall i = \overline{1, N}.$$

For each data vector, determine the closest center from χ_c.

$$C_i = \operatorname{argmin}_{X \in \chi_c} L(X, A_i) \forall i = \overline{1, N}.$$

Calculate distances from each data vector to the second nearest element (center) in χ_c.

$$D_i = \min_{Y \in (\chi_c \setminus \{C_i\})} L(Y, A_i).$$

3: For each $X \in \chi_c$, calculate $\delta_X = F(\chi_c \setminus \{X\}) = \sum_{i:C_i]X} (D_i - d_i)$.

4.1: Calculate n_δ.

$$n_\delta = \max\{[(|\chi_c| - p) * \sigma_e], 1\}.$$

Sort values of δ_X and choose a subset $\chi_{elim} = \{X_1, \dots, X_{n_\delta}\} \subset \chi_c$ from n_δ of coordinates corresponding to minimum values of δ_X.

4.2: For each $j \in \{\overline{2, |\chi_{elim}|}\}$, if $\exists k \in \{\overline{1, j-1}\}: L(X_j, X_k) < L_{min}$ then remove X_j from χ_{elim}.

4.3: Store $\chi_c = \chi_c \setminus \chi_{elim}$.

4.4: Reallocate data vectors between closest centers.

$$C_i^* = \operatorname{argmin}_{X \in \chi_c} L(X, A_i) \forall i = \overline{1, N}.$$

4.5. For each $X \in \chi_c$, if $\exists i \in \{\overline{1, N}\}: C_i = X \; and \; C^*i \neq X$ then recalculate center X^* кластера $C_X^{clust} = \{A_i | C_i^* = X, i = \overline{1, N}\}$. Set $\chi_c = (\chi_c \setminus \{X^*\}) \cup \{X\}$.

5: Go to Step 2.

Experimentally, authors [12] determine optimal values of parameters σ_e=0.25 and $L_{min} = \min_{X \in \chi_c}\{\max\{L(X, X_j), L(X, X_k)\}\}$.

New algorithm

Most algorithms such as ALA procedure or rather efficient genetic algorithm with recombination of fixed length center sets [21] require given value of the clusters quantity *p*. Other algorithms such as *X*-means [22] choose the best value of clusters number *p* with use of a special criterion. The adequacy of such criteria is a separate complex problem. Below, we propose a simple modification of the greedy heuristic: after reaching the quantity of clusters *p*, the process of elimination of the centers from the interim solution does not stop and algorithm fixes the values of the objective function for each value of centers quantity. Thus, we can obtain solutions for series of problems with $p = \overline{2, p_{max}}$.. However, the maximum value of clusters quantity p_{max} must be given.

The algorithm below is a combination of such greedy agglomerative heuristic with the genetic algorithm. However, such heuristic can be used with other global search strategies [23]

Algorithm 4. GA with greedy heuristic for solving series of problems with $p = \overline{2, p_{max}}$.

1. Initialization of a population of N_{pop} individuals. Each individual is a set of p_{max} centers (we denote it X and X_i is the ith element of this set). Set $F_{new,j}= +\infty$ for each $j = \overline{1, N_{pop}}$. Initialize the arrays of the objective function values $F^*_k = +\infty$ and best folutions $X^*_k=\{\}$ for each $k = \overline{2, p_{max}}$.

2. Select randomly $j_1, j_2 \in [1, N], j_1 \neq j_2$

3. $X_{new} = X_{j_1} \cup X_{j_2}$

4. While $|X_{new}| > p_{max}$:

4.1. Select an element j such that its elimination results in minimum increase of the objective function: $j = \arg\min_{j \in \chi_{\text{new}}} F(\chi_{\text{new}} \backslash \{j\})$

4.2. $X_{new} = X_{new} \backslash \{j\}$. Continue iterations 4.

5. Set $F_{new} = 0$; $X^* = X_{new}$.

6. While $|X_{new}| > 2$:

6.1. Set $F_{new} = F_{new} + f(X_{new})$; $k=|X_{new}|$; $F_k= f(X_{new})$; if $F_k < F^*_k$ then set $F^*_k = F_k$;

6.2. Perform Steps 4.1 and 4.2 for X_{new}. Continue iterations 6.

7. Choose j_3 using tournament selection by value of $F_{new,j}$. Set $F_{j_3} = F_{new}$; $X_{j_3} = X^*$, $F_{new,j_3} = F_{new}$.

8. Check the stop conditions, go to 2.

Computational experiments

Computational results are shown in Table 1. For the continuous p-median problems, comparatively precise results are achieved when the quantity of clusters is large. Having decreased this quantity p, we obtain comparatively worse results. Thus, new algorithm is efficient for : $p = \overline{p_{max}, \lceil p_{max}/3 \rceil}$. After reaching the minimum number of clusters $\lceil p_{max}/3 \rceil$, the algorithm must be restarted for obtaining results with less number of clusters.

For the EEE components classification problems, it is usually enough to solve the problems with p∈{2..10}. Such problems can be successfully solved by a single run of the new algorithm.

Conclusion

New genetic algorithm can be efficiently used for solving k-means and p-median problems which arise during the process of EEE components classification by homogeneous production batches. Computational experiments demonstrate the preciseness of the new algorithm and stability of its results (minimum standard deviation) which is very important for technical problems with high price of an error.

Table 1. Computational results for the p-median and k-means problems

Data set and its parameters	*p* and distance metric	Algorithm (see comments)	Time, sec..	Average result	Std. deviation
Test results of the electronic chip 1526TL1,	*p*=14, l_2^2	ALA multistart	15	150,124869801	0,384203928
		Sheng, Liu+ALA	15	149,954679652	0,172789313
		New algorithm	15 (for all problems)	149,78736565*	0,03157532*
N=1234, d=120 (in case of new algorithm,	*p*=10, l_2^2	ALA multistart	15	198,375350991	0,018643710
		Sheng, Liu+ALA	15	198,377650812	0,024878118
		New algorithm	15 (for all problems)	198,35974703*	$2\cdot10^{-14}$*
p∈{2..20})	*p*=6, l_2^2	ALA multistart	15	362,70701636*	0*
		Sheng, Liu+ALA	15	362,70401636*	0*
		New algorithm	15 (for all problems)	362,704051312	0*
UCI Mopsi Joensuu, N=6014, d=2,.	*p*=10, l_2	ALA multistart	15	359,680203232	3,964320582
		Sheng, Liu+ALA	15	359,545250068	2,526439494
		New algorithm	15 (for all problems)	359,41046080*	0,177992934*
(in case of new algorithm , p∈{2..20})	*p*=4, l_2	ALA multistart	15	596,825210394	0,000000442
		Sheng, Liu+ALA	15	596,82520843*	0,000000388
		New algorithm	15 (for all problems)	596,825283111	0*
BIRCH-3, N=100000, d=2,	*p*=100, l_2^2	ALA multistart	30	$3,7513245\cdot10^{13}$	116786778766
		New algorithm	30 (for all problems)	$3,740432\cdot10^{13}$*	21699776156*
(in case of new algorithm,	*p*=50, l_2^2	ALA multistart	30	$9,0099578\cdot10^{13}$	9545892119
		New algorithm	30 (for all problems)	$8,902789\cdot10^{13}$*	0*
p∈{2..110})	*p*=20, l_2^2	ALA multistart	30	$3,303278\cdot10^{14}$*	0*
		New algorithm	30 (for all problems)	$3,3049972\cdot10^{14}$	0*

Comments:" ALA multistart" means multiple starts of the ALA procedure with random initial seeding, "Sheng,Liu+ALA" means running genetic algorithm with recombination of the fixed length subsets powered by the ALA local search [21], "New algorithm" means Algorithm 4. The best results are marked by "*".

References

[1] Kazakovtsev L A, Antamoshkin A N, Masich I S 2015 Fast Deterministic Algorithm for EEE Components Classification *IOP Conf. Series: Materials Science and Engineering*, Volume 94, article 012015, 10 pages. doi:10.1088/1757-899X/94/1/012015

[2] Fedosov V V, Orlov V I. 2011 Minimal necessary extent of examination of microelectronic products at inspection test stage *Izvestiya Vuzov. Priborostroenie*, Volume 54(4), pp. 62-68.

[3] Kharchenko V S, Yurchenko Yu B 2003 Rating of fault-tolerant onboard complexes frames at usage electronic components industry *Tekhnologiya I konstruirovanie v elektronnoy apparature*, 2, pp. 3-10.

[4] Kazakovtsev L A, Orlov V I, Stupina A A, Masich I S 2014 Problem of electronic components classifying *Vestnik SibGAU*, issue 4(56), pp. 55-61.
[5] Ackermann M R et al. 2012 StreamKM: A Clustering Algorithm for Data Streams *J. Exp. Algorithmics*, Volume 17, 2.4:2.1-2.30.
[6] Kanungo T et al. 1999 Computing nearest neighbors for moving points and applications to clustering *Proc.of the tenth annual ACM-SIAM symp. on Discrete algorithms (Society for Industrial and Applied Mathematics)*, pp 931-932.
[7] Kazakovtsev L A, Stupina A A, Orlov V I 2014 Modification of the genetic algorithm with greedy heuristic for continuous location and classification problems *Sistemy upravleniya i informatsionnye tekhnologii*, No. 2(56), pp. 31–34.
[8] Hamiter L 1991 The History of Space Quality EEE Parts in the United States *ESA Electronic Components Conf., ESTEC (Noordwijk, The Netherlands, 12–16 Nov 1990 ESA SP-313)*
[9] Kirkconnell C S et al. 2014 High Efficiency Digital Cooler Electronics for Aerospace Applications *Proc. SPIE 9070, Infrared Technology and Applications XL 90702Q (June 24, 2014)*
[10] Farahani R Z, Hekmatfar M (editors) 2009 *Facility Location: Concepts, Models, Algorithms and Case Studies* (Berlin Heidelberg: Springer-Verlag)
[11] Kazakovtsev L A, Stupina A A 2014 Fast Genetic Algorithm with Greedy Heuristic for p-Median and k-Means Problems *IEEE 2014 6th Int. Congress on Ultra Modern Telecommunications and Control Systems and Workshops ICUMT (St.-Petersburg)*, pp. 702-706.
[12] Kazakovtsev L A, Antamoshkin A N 2014 Genetic Algorithm wish Fast Greedy Heuristic for Clustering and Location Problems *Informatica*, Volume 38(3), pp. 229-240.
[13] Deza M M 2013 Metrics on Normed Structures *Encyclopedia of Distances* (Berlin Heidelberg:Springer), pp. 89-99. doi: 10.1007/978-3-642-30958-85.
[14] Weiszfeld E 1937 Sur le point sur lequel la somme des distances de n points donnes est minimum *Tohoku Mathematical Journal*, Volume 43(1), pp.335-386.
[15] Cooper L 1963 Location-allocation problem *Oper. Res.*, Volume 11, pp. 331-343.
[16] Arthur D, Vassilvitskii S 2007 k-Means++: the Advantages of Careful Seeding *Proc. of the eighteenth annual ACM-SIAM symp. on Discrete algorithms*, pp. 1027-1035.
[17] Mishra N, Oblinger D, Pitt L 2001 Sublinear time approximate clustering *12th SODA*, pp. 439-447.
[18] Sun Zh et al. 2014 A parallel clustering method combined information bottleneck theory and centroid-based clustering *The Journal of Supercomputing*, Volume 69(1), pp.452-467.
[19] Alp O, Erkut E, Drezner Z 2003 An Efficient Genetic Algorithm for the p-Median Problem *Annals of Operations Research*, Volume 122, pp.21-42.
[20] Neema M N, Maniruzzaman K M, Ohgai A 2011 New Genetic Algorithms Based Approaches to Continuous p-Median Problem *Netw. Spat. Econ.*, Volume 11, pp. 83-99.
[21] Sheng W A, Liu X 2004 Genetic k-medoids clustering algorithm *Journal of Heuristics*, Volume 12(6), pp. 447-466.
[22] Pelleg D, Moore A 2000 X-means: Extending k-means with efficient estimation of the number of clusters *Seventeenth Internat. Conf. on Machine Learning*, pp. 727–734.
[23] Kazakovtsev L A, Antamoshkin A N 2015 Greedy heuristic method for location problems *Vestnik SibGAU*, Volume 16(2), pp. 317-325.

Energy regeneration model of self-consistent field of electron beams into electric power*

B N Kazmin, D R Ryzhov, I V Trifanov, A A Snezhko, M V Savelyeva
Reshetnev Siberian State Aerospace University
31, "Krasnoyarskiy Rabochiy" prospect, Krasnoyarsk, 660037, Russia.

E-mail: sibgau-uks@mail.ru

Abstract. We consider physic-mathematical models of electric processes in electron beams, conversion of beam parameters into electric power values and their transformation into users' electric power grid (onboard spacecraft network). We perform computer simulation validating high energy efficiency of the studied processes to be applied in the electric power technology to produce the power as well as electric power plants and propulsion installation in the spacecraft.

Energy regeneration model of electron (ion) beams greatly determines energy efficiency of electric power technology [3], where the beams are generated with electric arc plasma guns and electron guns with electric, magnetic and electromagnetic lenses by ionising the power fluid with electric arc and forming beams of the charged plasma [1, 2].

Self-consistent field energy [4, 5] generated by interaction of electric charges of electron (ion) beams is transformed and used in many technological processes: generation, amplification and transformation of microwave oscillation, as well as it is applied for wakefield acceleration of the charged particles in colliders [6] and in other processes [1].

To regenerate energy of the charged plasma beams into electric energy their kinetic energy of self-consistent field [3, 6] is mainly used. Potential energy of self-consistent field is practically not applied. Using potential energy significantly increases energy efficiency of electric power technologies, as the amount of potential energy in the charged plasma beams is practically equal to the amount of kinetic energy [1, 2] according to law of conservation of energy.

The chain diagram of the electric power plant to realise the proposed model of energy regeneration of self-consistent field of electron beams into electric power is in Figure 1.

The plant contains controlled voltage switch converters (CVSC) 1 transforming ac voltage into dc voltage, the last energises electrodes of electric arcs of electric guns (EG) 2. Electric arc anode (Aa) is connected with a «+» of CVSC terminal, cathode is connected with «–» to electric plant earth. AC voltage U_A of cyclic output frequency ω_p of users' electric power grid (UEPG) (onboard power system of a vehicle) energises output anodes of electric guns. Electric guns generate electron beams (EB) 3 pulsating with ω_p frequency.

Working spaces (WS) 4 are made of electron conductor material, therefore double electric layer (DEL) appears on their inner surface, it generates radial component of electric field $\mathbf{E}_r$. The radial component field density is $10^2 \ldots 10^3$ times higher than EG 3 field density, therefore, it compresses an electron beam transversally. Decelerating electrodes (DE) 5 look

* The results were obtained within the project No. 72 of the state task No. 2014/211 of the Ministry of Education and Science of the Russian Federation.

like Faraday cup [7] transforming electron beam charge 3 into equivalent electric current that generates a direct axis component of electric field $\mathbf{E}_l$ on decelerating circuit resistance to decelerate electron beam 3. Power transformer – converter (PTC) 6 in the primary circuit has got inductive windings L_1 and L_2, inductive circuit L_2 with condenser C_p is aligned to output frequency ω_p resonantly. PTC 6 secondary circuit is connected with users' electric power grid via electric power meter Wh of overcurrent type. Generating capacitance (GC) working at cyclic frequency ω_p supplies the energy to the users' electric power grid.

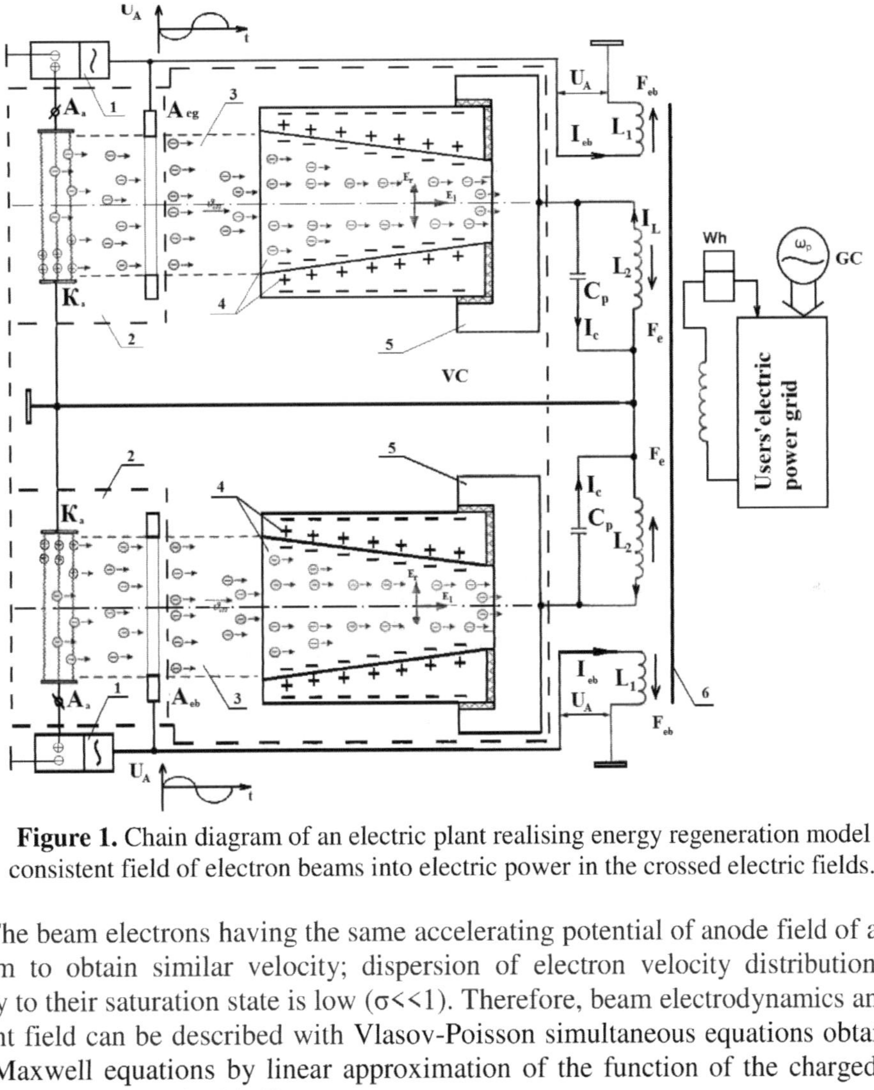

Figure 1. Chain diagram of an electric plant realising energy regeneration model of self-consistent field of electron beams into electric power in the crossed electric fields.

The beam electrons having the same accelerating potential of anode field of an electric gun seem to obtain similar velocity; dispersion of electron velocity distribution function relatively to their saturation state is low (σ<<1). Therefore, beam electrodynamics and its self-consistent field can be described with Vlasov-Poisson simultaneous equations obtained from Vlasov-Maxwell equations by linear approximation of the function of the charged particles distribution due to velocity [4, 5].

Vlasov-Poisson equations are simultaneous equations for every component of an electron beam. We consider a beam within nonrelativistic limit:

$$\frac{\partial f_\alpha}{\partial t} + \boldsymbol{\upsilon}\frac{\partial f_\alpha}{\partial \mathbf{x}} + \frac{q_\alpha \mathbf{E}}{m_\alpha}\cdot\frac{\partial f_\alpha}{\partial \boldsymbol{\upsilon}} = 0, \qquad (1)$$

Poisson equation is also applied ($\nabla \cdot \mathbf{E} = -\nabla^2 \varphi = \rho / \varepsilon$), as well as differential form of Poisson equation [9] for a self-consistent electric field ($div\mathbf{E} = -divgrad\varphi = \nabla^2 \mathbf{U} = \rho / \varepsilon$), Laplace equation in Cartesian axials ($\nabla^2 \mathbf{U} = \partial^2 \mathbf{U} / \partial x^2 + \partial^2 \mathbf{U} / \partial y^2 + \partial^2 \mathbf{U} / \partial z^2 = 0$).

The solution of the equations [8] for sinusoidal distribution of electron beam charges results in distribution of a potential $\varphi(\mathbf{x},t)$ of a self-consistent electric field of an electron beam.

$$\varphi(\mathbf{x}, \mathrm{t}) = \rho / 4\pi\varepsilon r \int_V dV = (\rho V / 4\pi\varepsilon r) \cdot \exp(j\omega_p t) \qquad (2)$$

Where: q_α- electric charge of an α component and m_α - mass of particles of an α component of an electron beam, $\mathbf{E}$ $(\mathbf{x},t)$- self-consistent electric field vector; $\varphi(\mathbf{x},t)$ - self-consistent electric field potential and ρ_a – charge density of an α component in an electron beam; $\rho V= q_a$ – electric charge of an α component of an electron beam with V volume; r – vector-range among electrons of a beam; ω_p – cyclic operation frequency modulating electron beam within an electron gun of electric arc plasmatron [1].

Potential distribution of a self-consistent electric field results in adequate intensity distribution of a self-consistent electric field, capacity and energy of an electron beam [4, 5, 8]:

$$\mathbf{E}(\mathbf{x},t) = -grad\varphi(\mathbf{x},t)\,;\; W_q = q_a\, \varphi(\mathbf{x},t) = q_a^2/4\pi\varepsilon r_e;\; S_q = W_q \cdot f_p\,; \qquad (3)$$

where: W_q and S_q – capacity and energy accordingly, they are generated by electron beam charges; f_p $=\omega_p/2\pi$ – operational frequency of anode field of electron gun modulating an electron beam.

The chain diagram of the electric plant is presented by Fig. 1. The diagram demonstrates simulation of the processes generating electron beams and transforming their energy into electric power parameters.

Ionising working fluid by electric arc plasmatron in the anode field of an electron gun removes electrons out of electric arc area gathering electrons into a beam. The beam is modulated by the operating frequency ω_p, that directs to a working space of a polarisable electrode. Electric power consumption of the process is:

$$W_{cons} = q_{eb}\,(U_{arc} + U_A) = q_{eb}\, U_A(1 + K_{W\,arc}); \qquad S_{cons} = I_A\, U_A(1 + K_{S\,arc}); \qquad (4)$$

where: q_{eb} – electric charge of an electron beam; U_{arc} – operational voltage of an arc (approximately 50 V); U_A – voltage of an output anode of an electron gun, it can change within a range of 500V up to 5000V; $K_{W\,arc}$ = 2,5 $U_{arc}/$ U_A = $K_{S\,arc}$ – consumption coefficient of energy or capacity, accordingly, to an electric arc relatively to capacity consumption of anode field of a gun 2.

Potential energy of electron range of self-consistent field generated by Coulomb and Lorentz orthogonal forces within an electron beam 3, in the first approximation, is the same, as a beam 3 is in electrodynamic balance under the impact of these forces, and a total beam energy 3 and its capacity equal to the sum of their effective values [2, 8]:

$$W_{qeb} = (W_{К}^2 + W_{Л}^2)^{1/2} \approx \sqrt{2}\, q_{eп}^2/\varepsilon r_e;\; S_{qeп} = W_{qeп} f_e; \qquad (5)$$

where: $W_C = F_C\, r_e = (q_{eb}^2/\varepsilon r_e^2)\, r_e = q_{eb}^2/\varepsilon r_e$ – energy component of self-consistent field of beam 3, generated by Coulomb forces; $W_L = F_L\, r_e = q_{eb}\, [\boldsymbol{v}_e \times \boldsymbol{B}_e]\, r_e \approx q_{eb}^2/\varepsilon r_e$ –energy component of beam 3, generated by Lorentz forces; $\boldsymbol{B}_e = \mu e v_e/2\pi r_e$ –self-consistent field density generated by an electron of beam 3 in the nearby electron.

Energy of the charged particle range and their kinetic energy in the electron beam generated by anode field of the electron gun is

$$W_{eb} = W_{qeb} + W_k = \sqrt{2}\, q_{eb}^2/\varepsilon r_e + N_{eb} m_e v_e/2 = 2\, q_{eb} U_A; \quad (5a)$$

where: $W_{qeb} = \sqrt{2} q_{eb}^2/\varepsilon r_e = q_{eb} U_A$ – potential energy of electron range in a beam generated by an electron gun; ε – dielectric permittivity in the electron beam environment; N_{eb} – the amount of electrons in a beam having potential of anode field U_A of an electron gun is characterised by quadratic dependence on electron beam charge. Therefore, there is an aerie of electric charge amount within an electron beam, where the energy of self-consistent field can exceed consumption of energy and capacity to generate an electron beam [9].

Energy consumption of anode field of electron gun to acceleration, mass transfer of electrons constituting a beam with kinetic energy [2] is:

$$W_A = q_{eb} U_A = N_{eb} e U_A \leftrightarrow N_{eb} m_e v_e^2/2 = W_k; \quad (6)$$

In anode field of electron gun electrons get the field energy and accelerate increasing kinetic energy. In decelerating field electrons decrease velocity and give kinetic energy to the field increasing the potential of decreasing field:

$$\Delta U \leftrightarrow m_e(\Delta v_e)^2/2e; \quad (6a)$$

where: ΔU – voltage (potential) change of field accelerating or decelerating the beam; Δv_e – corresponding change of electron velocity within a beam.

The expressions (5, 5a) show the consumption of electric power and capacity to form an electron beam is proportional to the beam electric charge, and energy of self-consistent field of an electron beam is described by the expressions (6, 6a) formed by the dynamic balance of orthogonal forces of Coulomb and Lorentz.

It is necessary to generate electron beams with the adequate amount of (generating) q_G electric charge and generating I_G electric current of a beam for the proposed model of processes of energy regeneration of electron beams to be both an electric power source and transformer of one electric energy form to another. The minimum generating charge and current power of an electron beam [9] are

$$q_{Gmin} \geq W_{qeb} / W_{A\,cons} = (\varepsilon e l_A U_A)^{1/2}; \quad (7)$$

$$I_{Gmin} = q_{Gmin} v_e / r_e = \varepsilon(2e/m_e)^{1/2}\, U_A^{3/2}. \quad (8)$$

Generating electric power requires following condition: $W_{qeb} / W_{cons} > 1$.

According to the condition electric charge of an electron beam should be not less than $W_k = N_{eb} m_e v_e/2 = q_{eb} U_A$ – kinetic energy of an electron beam,

$$q_{Geb} \geq (\varepsilon e l_A U_A)^{1/2}\,(1 + К_{WД}). \quad (9)$$

Respective to that, the electric current intensity of an electron beam generating electric power [10] is

$$I_{Geb} = \varepsilon(2e/m_e)^{1/2}\, U_A^{3/2}\,(1 + К_{SД}) \quad (10)$$

Minimum capacity of an electron beam in generation mode is

$$S_{G\,min} = I_{Geb} U_A \quad (11)$$

Minimum coefficient of electron beam capacity regeneration in generation mode is

$$K_{RS\,min} = S_{G\,min} / S_{cons} \quad (12)$$

Regeneration coefficient of electron beam capacity an operation mode of generation is

$$К_{RSp} = S_{G\,arc}/ S_{cons} = U_A \varepsilon (2e/m_e)^{1/2}\,(U_A(1 + К_{S\,arc}))^{3/2} / I_A U_A (1 + К_{S\,arc}) \quad (12)$$

Under the coefficient the electric current intensity of an electron beam I_{Geb} is increased taking into account the power consumption to the electric arc.

The results of computer simulation are presented by figure 2. The simulation is performed with software Delphi 7 [10], that allows to programme and research the presented above physic-mathematical models of electron beams and regeneration processes of their energy into the equivalent electric power transformed into the network of 50 Hz.

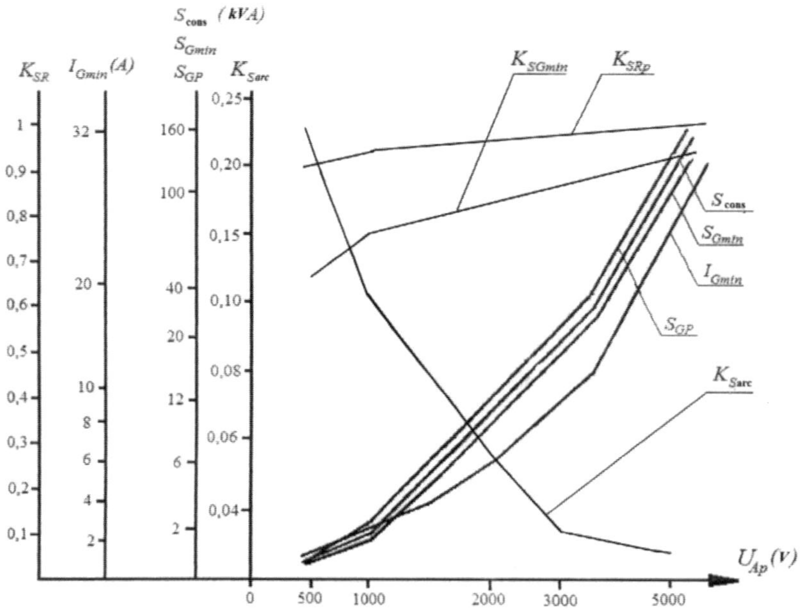

Figure 2. Results of simulating processes to generate electron beams within the crossed electric field and their energy regeneration into electric power transformed into users' electric power grid (onboard flying vehicle network):

I_{Gmin} (A) – electric current intensity in generation mode; S_{cons}(kVA) – power used to generate an electron beam; $S_{Gmin} = I_{Gmin}\ U_{Ap}$ (kVA) – minimum power in generation mode; S_{Gp} (kVA) – power in operation mode of generation; $K_{SGmin} = S_{Gmin} / S_{cons}$ – minimum coefficient power regeneration in generation mode; $K_{RSp} = S_{G\,arc} / S_{cons}$ - coefficient of power regeneration in operation mode of generation, expression (13); U_{Ap} (V) – operation voltage at the output anode of electron gun.

Therefore, decelerating with simultaneous keeping an electron beam in the compressed condition by "the crossed electric field" applied according to the proposed regeneration model of kinetic and potential energy of electron range of self-consistent field of electron beam into electric power transformed to the users' electric power grid (onboard flying vehicle network), it is possible to produce electric power with high efficiency; the electric power is equivalent to electron beam energy, where the beam is in the compressed condition.

The proposed model can be applied to produce electric power [11].

Summing it up, we could point out:

Firstly, it is technically simpler and more energy efficient to apply the models of the crossed electric field of electric arc, ionising working fluid, with anode field of an electron gun generating an electron beam and performing its mass transfer. It gives the chance to generate electron beams with current of higher intensity in accordance with the conventional generating electron beams in electric and magnetic fields.

Secondly, it significantly increases energy efficiency of the proposed model to generate electric power if we use the model of the cross electric field with double electric layer keeping an electron beam in the compressed condition with the field decelerating the beam. The field is generated by an electron beam to regenerate potential and kinetic energy of electron beam to electric power transformed to the users' electric power grid (onboard flying vehicle network).

Thirdly, double electric layer is generated due to electrophysical properties of materials, therefore its application to keep an electron beam in the compressed condition is not energy consuming, consequently, it increases energy efficiency of the proposed model to generate electric power.

References

[1] Enciklopediya nizkotemperaturnoj plazmy (Encyclopedia of Low-Temperature Plasma)/4 vol. Edited by V.E-Fortov. Moscow: Nauka, 2000

[2] Bolshoj Enciklopedicheskij Slovar. Fizika (Big Encyclopaedic Dictionary. Physics / Editor-in-Chief A.M. Prohorov. Moscow: Ross. Encyclopedia, 1998 – 994 p.

[3] Smirnov V.M., Perelygin S. F., Kondakov V. V. Ocenka perspektiv preobrazovaniya ehnergii plazmy v ehlektricheskuyu v zamknutyh magnitnyh lovushkah. Godovoj otchet po NIR kafedry fiziki plazmy za 2008 g. Moscow: NIYAU MIFI, 2009. - s. 5-6. (Estimating the prospects of plasma energy conversion into electric energy in closed magnetic hooks/Annual report on the research by the Department of plasma physics for 2008. Moscow: National Research Nuclear University (Moscow Engineering Physics Institute), 2009. - p. 5 - 6.)

[4] Vlasov A.A. Teoriya vibracionnyh svojstv ehlektronnogo gaza I ee prilozheniya (Theory of vibration properties of electron gas and its application) Moscow: Moscow State University – 1945 – issue 75, book 2, part 1.

[5] Vlasov A.A. Statisticheskie Funkcii Raspredeleniya (Statistical distribution function. 2 Ed., corr. Vlasov A.A. 2014. – 360 p.)

[6] Aleksandrov A.F., Kuzelev M.V.. Teoreticheskaya plazmennaya ehlektrotekhnika. Uchebnoe posobie. MGU imeni M.V. Lomonosova, Fizicheskij fakul'tet, 2011. - 167 s. (Theoretical plasma electrical engineering. Tutorial. Moscow: Lomonosov Moscow State University, Faculty of Physics, 2011. - 167 p.)

[7] Yandex. Ru/ cilindr Faradeya (Faraday Cup)/ stu.scask.ru/book.ar1.php?id=26.

[8] Shimoni K. Teoreticheskaya elektrotekhnika / M., «Mir», 1964. - 773 s. (Shimoni K. Theoretical electrical engineering/Moscow: Mir, 1964. - 773 p.)

[9] Kazmin B.N., Trifanov I.V. Ob ehlektronnom generatore ehlektroehnergii // B.N. Kazmin, I.V. Trifanov /Vestnik SibGAU. Vyp. 1(34). 2011. - s. 25–28. (About electronic generator of electric power // B.N. Kazmin, I.V. Trifanov / Vestnik SibSAU. Issue 1(34), 2011. - p. 25–28.

[10] Arhangel'skij A.Ja. Delphi 7 : spravoch. posobie. Moscow : Binom-Press, 2004. – 1024 s. (Archangelsky A.Ya. Delphi 7: resource book. Moscow: Binom-press, 2004. – 1024 p.)

[11] Patent № 2505915 RF, МПК H02N 3/00./ Elektronnyj generator elektroehnergii (Electron electric power generator) / B.N. Kazmin, Trifanov I.V., Oborina L.I., Dubova E.D., Sterechov I.V. Published 27.01.2014. Bulletin № 3.

The model of electrophysical processes increasing effectiveness of electric power technology based on electron beams *

B N Kazmin, I V Trifanov, D R Ryzhov, M V Savelyeva
Reshetnev Siberian State Aerospace University
31, "Krasnoyarskiy Rabochiy" prospect, Krasnoyarsk, 660037, Russia.

E-mail: sibgau-uks@mail.ru

Abstract. The research focuses on electrophysical model of processes increasing energy efficiency of electric power technology based on electron beams that are generated by crossed electric field. The energy of beams is transformed with compression and simultaneous deceleration with crossed electric field into electric power to transfer into users' electric power grid. We perform computer modeling of the processes to confirm the energy efficiency of the proposed model.

We suggest examining techniques and means to generate electron (ion) beams and transform their energy into electric power, and obtaining models to increase energy efficiency of these electric power processes.

It is possible to specify electromagnetic processes in electron (ion) beams of low-energy plasma, generated in plasmatrons with electron (ion) guns with electric and magnetic lenses [1, 2], applying Maxwell-Lorentz electromagnetic field equation [3] by means of electric currents in electron beams ($I_{eb} = I_c + I_{cond} + I_{sh}$): convection I_c, conductivity I_{cond} and shift I_{sh}. Vlasov-Poisson simultaneous equations can describe electrodynamics of plasma processes obtained from Vlasov-Maxwell simultaneous equations [4, 5].

Electron (ion) beams with high energy density are widely used in various energetic technologies both in aerospace industry and in many other industries in Russian economy [1, 6, 7, 8].

The technology "to generate electric power" [6, 7] is known, due to it an electric arc ionizes power fluid and generates electron beams with anode bed of an electron gun; anode bed crosses with arc electric field. A model to generate electron beams in "the crossed electric fields" is less electric power consuming and technically less complicated in comparison with conventional methods of generating electron beams in electric and magnetic fields. [1, 2].

To increase energy efficiency of electric power technology transforming energy generated in crossed electric fields of electron beams into electric capacitance is realized with crossed electric fields.

It is possible to realise a model of electric physical processes increasing energy efficiency of electric power technology to produce electric power with generating electron beams in the crossed electric fields and transforming energy of electric beams into electric capacitance in the crossed electric fields while using electric plant.

* The results were obtained within the project No. 72 of the state task No. 2014/211 of the Ministry of Education and Science of the Russian Federation.

IOP Conf. Series: Materials Science and Engineering **122** (2016) 012013 doi:10.1088/1757-899X/122/1/012013

Fig. 1 presents a functional diagram of the electric plant. The plant contains controlled voltage switch converters (CVSC) 1 transforming ac voltage into dc voltage, the last energises electrodes of electric arcs of electric guns (EG) 2. Electric arc anode (Aa) is connected with a «+» of CVSC terminal, cathode is connected with «—» to electric plant earth. AC voltage U_A of cyclic output frequency ω_p of users' electric power grid (UEPG) (onboard power system of a vehicle) energises output anodes of electric guns. Electric guns generate electron beams (EB) 3 pulsating with ω_p frequency. Working spaces (WS) 4 are made of electron conductor material, therefore double electric layer (DEL) appears on their inner surface, it generates radial component of electric field $\mathbf{E}_r$. The radial component field density is 10^2... 10^3 times higher than EG 3 field density, therefore, it compresses an electron beam transversally. Decelerating electrodes (DE) 5 look like Faraday cup [5] transforming electron beam charge 3 into equivalent electric current that generates a direct axis component of electric field $\mathbf{E}_l$ on decelerating circuit resistance. The component decelerates electron beam 3. Power transformer – converter (PTC) 6 in the primary circuit has got inductive windings L_1 and L_2, inductive circuit L_2 is aligned to output frequency ω_p resonantly. PTC 6 secondary circuit is connected with electric power consumer line via electric power meter Wh of overcurrent type. Generating capacitance (GC) working at cyclic frequency ω_p supplies the energy to the users' electric power grid.

The electric plant according to the proposed technique to generate electric power operates in the following way. The necessary voltage is fed to Aa and Ka electrodes of an electric arc with CVSC 1; power fluid filling a vacuum chamber (VC) is ionised under the ark influence. The impact of half-wave voltage U_A fed to output anode A_{eg} of an electron gun 2 generates anode electric field crossed with the electric arc field. Electron beams 3 are generated in this "crossed field" and they direct to a work space 4, where "the crossed electric field"[1] impacts on electron beam 3.

[1)] We introduce a new term – "the crossed electric field", by analogy with the term "the crossed electric and magnetic field" used in science-technical literature; here we could also apply terms "orthogonal electric field" or "normal" and "tangential component" of electric field but not for the proposed engineering solution. The fields can be non-orthogonal and working frequencies of the crossed fields are different in the proposed model. Generating electron beams in the crossed electric field are unsophisticated and more energy-efficient than in the "the crossed electric and magnetic field" [1, 2, 15].

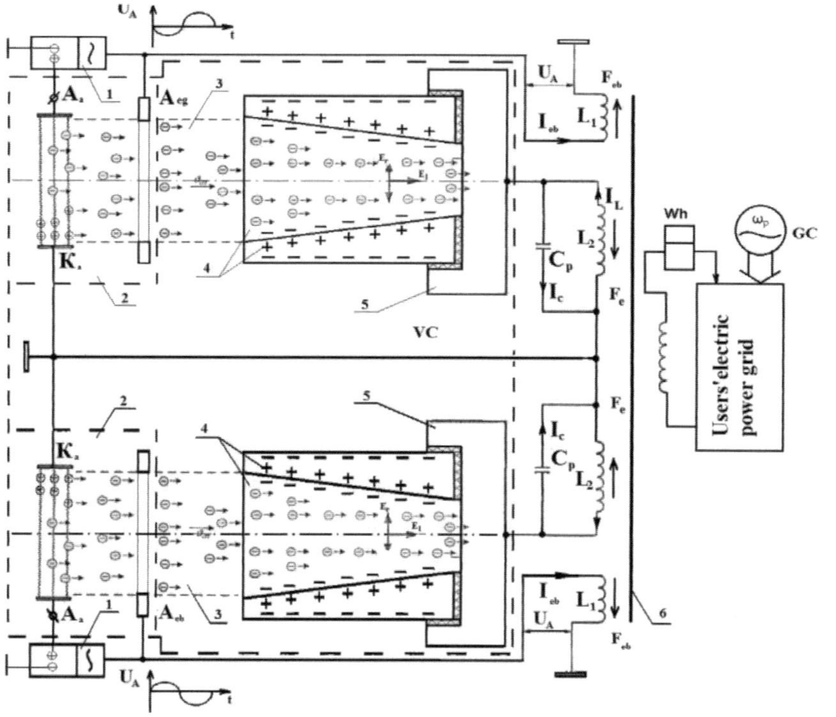

Figure 1. Functional diagram of an electric plant realising a model of electrophysical processes of electric power technology based on electron beams in the crossed electric fields.

During the period of time cations of the ionised power fluid come to the arc cathode (C_A) charging it positively. The cations have a mass of thousand times bigger than electrons have, therefore they do not practically change their trajectory of movement within an electric arc.

Radial component $\mathbf{E}_r$ of the crossed field generated by double electric layer in WS 4, compresses an electron beam transversely due to taper pencil along the beam movement. Beam compression coefficient 3 in WS 4 is

$$C_{cb} = d_{input}/d_{output}, \qquad (1)$$

where: d_{input} and d_{output} are diameters of input and output inlets in WS 4 accordingly.

Electron beam compression can be realized with electric, magnetic, electro-magnetic "quadrupole electron lenses" allowing to control electron beam compression 3 in an electric plant, though it requires to consume additional energy. Therefore, our invention proposes an energy-efficient technical solution: to compress electron beam with the field of double electric layer generated due to electrophysical properties of materials without additional energy consumption [2].

Compressing electron beam 3 results in electron approach to each other and increase in charge density in a beam

$$\rho_{eb} = \rho_e C_{cb} \approx C_{cb} 3e/4\pi r_e^3, \qquad (2)$$

where: $e=1{,}6\cdot 10^{-19}$ C – electron charge; r_e – an average distance among electrons in a beam; $4\pi r_e^3/3$ – volume an electron has got in a ray; the increase in energy of electron interaction in a beam – potential energy in an electron beam

$$W_{ep}=q_{ep}^2/\varepsilon r_e = \rho_{ep}^2 V_{ep}^2/\varepsilon r_e\,, \qquad (3)$$

IOP Conf. Series: Materials Science and Engineering **122** (2016) 012013 doi:10.1088/1757-899X/122/1/012013

where: $q_{eb} = \rho_{eb} V_{eb}$ – the total charge of electrons in a beam; $V_{eb} = s_{eb} l_{eb}$ – volume, beam area and length of electron beam, accordingly; ε – dielectric permittivity of medium in an electron beam.

Increasing charge density consequently results in increasing electric current intensity and beam power 3

$$I_{eb}=\rho_{eb}\upsilon_e s_{eb}, \quad (4)$$

where: $\upsilon_e = (2eU_A/m_e)^{1/2}$ – electron velocity in a beam, s_{eb} – an area of beam radial section.

Potential power generated by electric charges of an electron beam compressed in WS 4.

$$S_{eb} = I_{eb} U_q = I_{eb} W_q / q_{eb} \quad (5)$$

Direct axis component $\mathbf{E}_l$ of the crossed field in WS 4 is generated by equivalent electron current obtained from charges q_{eb} of a beam 3 with Faraday cup [8] in an electrode 5. Electron current running along decelerating electrode circuit resistance 5 generates decelerating potential, that decelerates electron ray transforming potential energy of electron interaction compressed in WS 4 of an electron beam into its kinetic energy, equivalent power to the electric user power grid via L_2 PTC 6. The circuit load of decelerating electrode is tank resistance generated by inductance L_2 and condenser C_p. Current resonance ($I_L = I_C$) in a contour improves oscillation form decreasing energy loss to higher harmonic components and together compensates reactive power increasing power coefficient (cosφ) in the electric current practically up to "1". Decelerating potential of the component $\mathbf{E}_l$ of the crossed field in WS 4 is generated in antiresonant contour, magnetic flux vector is generated by inductance L_2

$$\mathbf{F_e} = I_{eb}\mu s_\mu \, exp(j\,\omega_p t\,)/l_\mu = \omega_p L_2 I_{eb}\, exp(j\,\omega_p t\,), \quad (6)$$

where: μ – magnetic inductive capacitance of coil flux guide material PTC 6; s_μ and l_μ – cross section and coil flux guide length PTC 6, accordingly; transformed potential power is obtained through L_2:

$$S_{ebL} = \omega_p L_2 I_{еп}^2. \quad (7)$$

Kinetic energy taken from electrons of a beam 3 while it is decelerated by the component $\mathbf{E}_l$ of field in WS 4 is also transformed via L_2 as kinetic power

$$S_{eb\upsilon L} = U_{eт}^2 / \omega_p L_2 = (m_e(\upsilon_e^2 - \upsilon_{eт}^2)/2)^2 / \omega_p L_2, \quad (8)$$

where: $U_{eт} = m_e(\upsilon_e^2 - \upsilon_{eт}^2)/2$ – decelerating electrode voltage 5 as the result of deceleration a beam 3 of the component $\mathbf{E}_l$ of the field in WS (РП) 4; $\upsilon_{eт}$ – electron velocity of a beam 3; the electrons belong to the decelerating electrode 5. We can consider equal both potential S_{eqL} and kinetic $S_{e\upsilon L}$ powers, generated by orthogonal Coulomb-Lorentz force in an electron beam 3 for the first approximation, as a beam 3 in electrodynamic balance under the force impact, and the total of the generation by the compressed beam 3 equals to the sum of effective ranges composing powers:

$$S_G = (S^2_{ebL} + S^2_{eb\upsilon L})^{1/2} = \sqrt{2}\, S_{ebL}. \quad (9)$$

The energy of electron ray generation producing the generation power S_G is:

$$W_G = \sqrt{2} W_q = \sqrt{2}\, q_{еп}^2 / \varepsilon r_e. \quad (10)$$

Output power necessary to generate an electron ray 3 with electron gun 2 is obtained from electric user power grid via inductance circuit L_l

$$S_{consump} = \omega_p L_1 I^2_{ЭП} = S_Д + S_A = U_A I_A (1 + К_{SД}), \quad (11)$$

where: $I_{ЭП} = I_A(1 + К_{SД})$ – the require current intensity for an electric gun; $S_Д$ and S_A - power output of electric arc and anode field of EG 2, respectively; $К_{SД} = 2{,}2U_Д/U_A$ – power output coefficient per electric arc in relation to power output of anode field of gun 2. The relevant magnetic flux within a coil flux guide of PTC 6 is generated to transfer current out of the grid.

$$\mathbf{F_{eb}} = I_{eb}\mu s_{\mu}/l_{\mu} \qquad (12)$$

The expressions (6 and 12) describing F_e and $\mathbf{F_{eg}}$ magnetic flux vectors reveal that vectors are performed by the dissimilar current $I_{еп}$ and $I_{ЭП}$, therefore, vectors $\mathbf{F_e}$ and F_{eb} are opposing to one another. Vector $\mathbf{F_{eb}}$ transferring power to EG (Эп) 2 is balanced by vector $\mathbf{\Phi_e}$ generating power. Generating power by the proposed electric plant requires vector $\mathbf{\Phi_e}$ of magnetic flux exceeds vector $\mathbf{F_{eb}}$ of magnetic flux:

$$\mathbf{F_e} / \mathbf{F_{eb}} \geq 1 \text{ или } I_{eb} = C_{сп}\, \rho_e \upsilon_e s_{еп} / I_{ЭП} = I_A(1 + K_{SД}) \geq 1. \qquad (13)$$

Satisfying the condition the electric power meter Wh of overcurrent type registers the amount of electric power produced by the plant while the current comes to the users' grid (onboard grid of a space vehicle).

The expression (13) results in necessary coefficient of electron ray compression to generate the required amount of power $S_{Gеп}$ by the electric plant and relevant coefficients to transform power generation K_{SG} and power of electron beam $K_{SП}$.

$$K_{сп} \geq I_A(1 + K_{SД}) / \rho_e \upsilon_e s_{еп} \qquad (14)$$

$$K_{SП} = K_{SG}/\sqrt{2} = S_{еп}/ S_{затр} = U_A K_{Cb}\, \rho_e \upsilon_e s_{eb} / U_A I_A(1 + K_{SД}) \qquad (15)$$

Electrons performing the work within the outer electric circuit under the impact of potential difference between decelerating electrode 5 and cathode $K_Д$ of electric gun 2 come to the cathode and recombine cations to atoms and molecules of working fluid under reionisation by electric arc between anode A_a and cathode C_a to start a new cycle of generating electric power and its transforming into users' grid (onboard grid of a space vehicle [9]).

The expressions (2, 3, 9, 10, 13, 14 and 15) demonstrates the generated power depends on electron ray compression coefficient C_{cb}, therefore changing C_{cb} controls power transformation coefficient and the amount of generated power according with the proposed energy production technique.

When changed automatically with frequency ω_p of polarity of half-wave voltage U_A on output anodes A_{eg} of electric gun 2, the different electron gun generates an electron beam 3 the described above processes happen where beam energy converts into electric power and further the different symmetrical branch of PTC 6 transforms it to the users' electric power grid. As a result we have full-wave transformation of electron beam (3) energy into electric power; its quality corresponds to the requirements of GOST 13109-97 [10]. The working fluid is not spent; there are no harmful wastes and emissions having negative impact on the environment, though they are typical to the power generating industry [11] burning fuel.

Fig. 2 presents the electric parameters obtained with computer modeling in Delphi [12, 13, 14, 16] of electrophysical processes of electric power technology based on electron beams in the crossed electric fields. The technology is realised with the electric power plant demonstrated by Fig. 1.

IOP Conf. Series: Materials Science and Engineering **122** (2016) 012013 doi:10.1088/1757-899X/122/1/012013

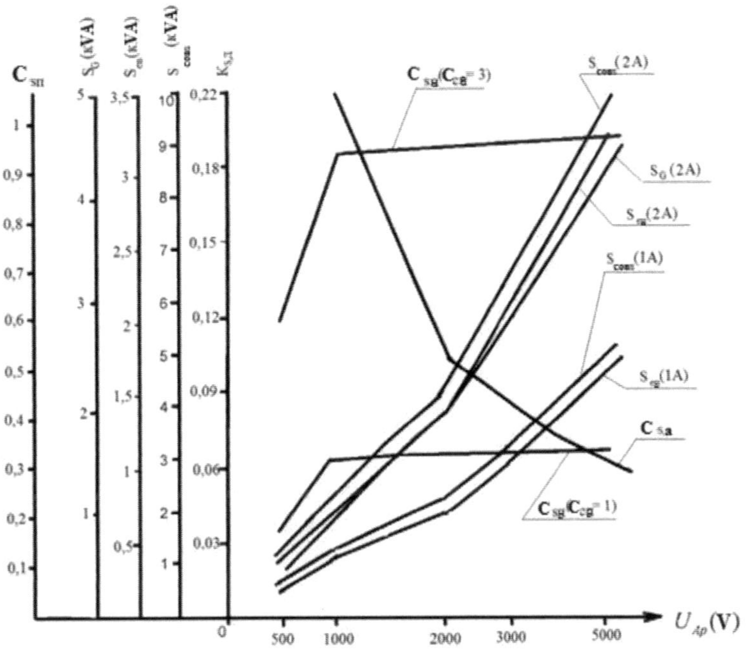

Figure 2. The results of modeling processes of electric power technology based on electron beams in the crossed electric field.

S_{cons}(1A) и S_{cons}(2A) - power consumption to generate an electron beam under the dynamic current of electron gun 1A and 2A accordingly;

$C_{SД}$ – coefficient of power consumption per electric arc relatively to power consumption in electric gun;

S_{eb}(1A) and S_{eb}(2A) – electric beam power under the dynamic current of electric gun 1A and 2A accordingly;

S_G(2A) – the generation power under the dynamic current of electron gun 2A;

C_{sb} – coefficient of electron beam power transformation.

Proportionally to the compression coefficient compression of electron beams 3 in WS 4 by the field of double electric layer not requiring additional electric energy consumption; it increases the energy efficiency of the proposed technology to produce electric energy.

Current resonant condition, compensating reactive power within the electric circuit decelerating the compressed electron beam 3 and converting equivalent power produced by the energy of the compressed beam, increases the electric circuit power coefficient up to «1». The power decelerates electron beam and transforms its energy to the users' electric power grid.

Therefore, deceleration and simultaneous electron beam compression applied according to the proposed model of electric power technology of energy production give the possibility to use both kinetic and potential energy of electron beam electric charges, consequently, to generate electric power with higher electric efficiency; the power is equal to the energy of the compressed electron beam. [15]

The proposed model of electrophysical processes increasing power generating technologies can be used for both electric power production [7] and aerospace engineering [8].

Summing it up we could state the following:

1. The proposed model to generate electron beams within "the crossed electric field" their energy regeneration within "the crossed electric field" into electric power transformed into users' electric power grid allows to increase energy efficiency of electric power technology based on electric beams.

2. Applying the crossed electric field of electric gun ionising power fluid together with anode field of electron gun generating electron beam with simultaneous electron mass transfer, enables to generate electron beams of higher intensity with higher energy efficiency technically simpler than the traditional technology based on electric and magnetic fields.

3. Compressing electron beam by the field of double electric layer not requiring additional energy consumption increases energy parameters of an electron beam and efficiency of the proposed model to generate electric power proportionally to compression coefficient.

Decelerating and simultaneous compressing an electron beam within the crossed electric field allows to transform both kinetic and potential energy of the interacted charges of an electron beam; therefore it transforms energy of the compressed electron beam into equivalent electric power with high energy efficiency to the users' electric power grid.

4. Current resonant condition within electric circuit decelerates the compressed electron beam and approaches the electric circuit power coefficient to «1» decelerating electron beam and transforming equivalent power to the users' electric power grid.

References

[1] Enciklopediya nizkotemperaturnoj plazmy (Encyclopedia of Low-Temperature Plasma)/4 vol. Edited by V.E-Fortov. Moscow: Nauka, 2000

[2] Bolshoj Enciklopedicheskij Slovar. Fizika (Big Encyclopaedic Dictionary. Physics / Editor-in-Chief A.M. Prohorov. Moscow: Ross. Encyclopedia, 1998

[3] Shimoni K. Teoreticheskaya elektrotekhnika / M., «Mir», 1964, 773 s. (Shimoni K. Theoretical electrical engineering/Moscow: Mir, 1964, - 773 p.)

[4] Vlasov A.A. Statisticheskie Funkcii Raspredeleniya (Statistical distribution function. 2 Ed., corr. Vlasov A.A. 2014. – 360 p.)

[5] Vlasov A.A. Teoriya vibracionnyh svojstv ehlektronnogo gaza I ee prilozheniya (Theory of vibration properties of electron gas and its application) Moscow: Moscow State University – 1945 – issue 75, book 2, part 1.

[6] Patent № 2262793 RF, МПК H02N 3/00. Sposob proizvodstva ehnergii (Method of producing energy)/ B.N. Kazmin. Published 20.10.2005. Bulletin № 29.

[7] Patent № 2505915 RF, МПК H02N 3/00./ Elektronnyj generator elektroehnergii (Electron electric power generator) / B.N. Kazmin, Trifanov I.V., Oborina L.I., Dubova E.D., Sterechov I.V. Published 27.01.2014. Bulletin № 3.

[8] Trifanov I.V. et al. Principy postroeniya ehlektroehnergeticheskih I ehlektrodinamicheskih tekhnologij kosmicheskih apparatov : monografiya (Principles of construction of electric and electrodynamic spacecraft: monograph) / I.V. Trifanov, B.N. Kazmin, D.R. Ryzhov, I.I. Homenko. Krasnoyarsk, Reshetnev Siberian state aerospace university, 2015 – 182 p.

[9] Yandex. Ru/ cilindr Faradeya (Faraday Cup)/ stu.scask.ru/book.ar1.php?id=26.

[10] Mezhgosudarstvennyj standart GOST 13109-97/Ehlektricheskaya ehnergiya. Sovmestimost' tekhnicheskih sredstv ehlektromagnitnaya. Normy kachestva ehlektricheskoj

ehnergii v sistemah ehlektrosnabzheniya obshchego naznacheniya// (vveden v dejstvie postanovleniem Gosstandarta RF 28.081998. № 338). (Interstate standard GOST 13109-97/electrical energy. Electromagnetic compatibility of technical equipment. Electric energy quality standards in power supply systems for general purpose//(introduced by the Decree of Gosstandart of the Russian Federation from August 28, 1998 No. 338).)

[11] Toplivno-energeticheskij kompleks Rossii // Informacionno-analiticheskij centr. Moscow: Energija, 2007. 478 s. (Fuel and energy complex of Russia//informational-analytical Centre. M.: energy, 2007. – 478p.)

[12] Arhangel'skij A.Ja. Delphi 7 : spravoch. Posobie. Moscow : Binom-Press, 2004. – 1024 s. (Archangelsky A.Ya. Delphi 7: resource book. Moscow: Binom-press, 2004. – 1024 p.)

[13] Glushakov S. V., Klevtsov A. L. Programmirovanie v srede Delphi 7.0. Moscow : Folio. 2003. – 415 s. (Glushakov C.V., Klevtsov A. L. Programming in Delphi 7.0. Moscow: Folio, 2003 – 415 p.)

[14] Suharev M. V. Osnovy Delphi. Professional'nyj podhod. M. : Nauka I tehnika, 2004. – 420 s. (Sukharev M. V. Delphi Basics. Professional approach. Moscow: Science and technology, 2004. – 420 p.)

[15] Ryzhov D.R., Kazmin B.N., Trifanov I.V. The model of plasma-electronic technology of producing electricity from electron beams / Ryzhov, D.R., Kazmin, B.N., Trifanov, I.V. // IOP Conference Series: Materials Science and Engineering, 2015.

[16] Kazmin B.N., Trifanov I.V., Ryzhov D.R., Savelyeva M.V. The alternative energy yield out of electron plasma flow/ B.N. Kazmin, I.V. Trifanov, D.R. Ryzhov, M.V. Savelyeva// International journal of applied and Fundamental research, 2013, №2 www.Science-sol.com/455-24/74, 10.11.2013.

IOP Conf. Series: Materials Science and Engineering **122** (2016) 012014 doi:10.1088/1757-899X/122/1/012014

On the question of the necessity of implementation of automatic control systems in timber industry

E R Khasanov, P V Zelenkov, M O Petrosyan, A V Murygin, V D Laptenor
Siberian State Aerospace University named after Academician MF Reshetnev
31 "KrasnoyarskiyRabochiy" prospect, Krasnoyarsk, 660037, Russia.

E-mail: xqish19@gmail.com

Abstract. The paper considers the necessity of implementation of automatic control systems on the level of forest farms management and timber industry. Main areas of activity, which currently subjected to automation, are revealed. Objectives, which solved by implementation of APCS, are identified.

Due to development of IT technologies automatic control systems are becoming increasingly more important in forest industry. These systems contain hardware and software, which allow to automate routine part of work for forestry specialists and timber industry specialists.

Forestry is a system of control. Forest resources act as a controlled object. Forest it is a complicated system. To control a complicated object, it is necessary to master modern approaches and tools. IT technologies and automatic systems are serving as management tools. Automation is defined as "use of technical equipment, mathematical methods and control systems that would partially or entirely relive people from direct participation in the process of obtaining, transformation, transmission and use of energy, materials and information", all of the following processes had received the name of "information technologies" now.[2]. Depending on level of automation, for example, process of gathering, processing and analyzing silvicultural information, systems can be divided on automated and automatic.

There are three automatic groups of systems in forestry:

1) Automation of scientific researches;

2) Computer-aided design;

3) Automation of forest industry.

At the present stage of technological progress, when in manufactory number of products doubles in fifteen years and their complexity doubles in ten years, when the risk of project solutions is very high, specialist engineers of forest industry are presented with high demands in the field of computer training

Forestry engineer should know:

• methods of mathematical statistics;
• basics of numerical methods;
• methods of mathematical modeling;
• basics of system analysis;
• automatic systems of scientific researches;
• automatic information systems, etc.

Furthermore, specialist of timber industry should be able to work on personal computers with automatic branch systems and control the manufacturing of finished product on the final stage. In manufacturing process product cost proceeds from production sphere to circulation sphere. [3].

Currently forest industry undergoes noticeable changes: companies' structures are changing, relatively small companies are becoming bigger and starting to unite in holdings, the processes of consolidation are taking place. With rise of volumes and manufacturing directions expansion appears a need for automation of company's management.

A lot of timber processing factories are automated in the area of accounting and reporting generation and costs, but currently this is not enough for effective management of company. [1,4,5]. Forest industry requirements for automation systems are getting higher: with further growth of business there is a need for a more efficient management of production processes, material and financial flows. The goal of IT technologies in company is the development of unified information space, which would cover all main objects of accounting. For staff APCS must become a useful everyday tool, which simplifies daily duties and free from routine job. For top managers APCS might become a flexible management tool in difficult, constantly changing circumstances. A tool, which allows to make more effective management decisions on account of improved quality of information provided. [6] In timber industry it is never easy to choose the right program part for development of such system. Due to dependence on business area, the need in control automation in logging manufactories differs from manufactories engaged in construction of wooden buildings. Herewith it is necessary to be clear about the list of tasks that must be addressed by the system, and focus on the market of software solutions.

Depending on company specifics, goals of automation could differ considerably. Thus, the need for automation of a small sawmill differ significantly from the complex needs of the wood processing enterprise.

Almost every wood processing company has standard set of goals, automation of which have different significance depending on company's caliber:

- operative accounting of production processes;
- maintenance of technological and engineering specifications;
- Capacity Planning;
- formation of sales and production plans;
- inventory control and procurement;
- operative financial control.

Control systems implementation allows reaching competitive advantages due to optimization of business processes and costs lowering. These systems were developed for prime cost management and further achieve competitive advantages.

Complex information systems, which take into account specifics of wood processing factories, include methods of planning and control. They allow to:

- adjust the amount of reserves;
- reduce the amount of unfinished production by scheduling production based on demand for the final products;
- evaluate the effectiveness of the execution of orders received on the basis of capacity of the enterprise;
- reduce costs and time spent on the production;
- monitor the actual performance results of each production unit and compare them with the plan;
- make quick adjustments in production plans;

- monitor the promptness of deliveries.

Thereby, periodical analysis of company's structure in APCS allows to watch over condition of production system, concentrate attention on specifics of main blocks, having in mind the increase of complex company's efficiency.

[1] FA Aklanov, Y. Nurgaleeva, OO Smirnov, ER Khasanov, MO Petrosyan. 2014 Rank overall effectiveness of the equipment with the support of decision-making on the maintenance and repair. *Reshetnev reading*. **T. 2**. № 18.
[2] VM Glushkov. Fundamentals of Paperless Informatics. 1987 2nd ed. P 552.
[4] Trahtengerts, EA 2012 Subjectivity in computer support management decisions. P 256.
[5] ER Khasanov, PV Melnikov OA Smirnov, KK Bahmareva. 2014. On the analysis of the efficiency of the utilities. *International Research Journal*. № 7-1 (26). P. 60.
[6] ER Khasanov, PV Melnikov. 2012. Information technology in assessing the efficiency of enterprises. *Collected materials of All-Russian Youth Conference "Information and telecommunication systems and technologies*. P. 82.

Study of a lithium-ion battery charge-discharge test unit characteristics*

E A Kopylov, E A Mizrah, A S Fedchenko, D K Lobanov[1]

[1] Siberian State Aerospace University named after academician M.F.Reshetnev, Krasnoyarsk, Russia

E-mail: alek.fedchenko@gmail.com

Abstract. The article describes the structure of a charge-discharge unit which allows to perform electrical, resource and thermal testing of several lithium-ion batteries simultaneously. The principle of operation of a one battery research channel (BRC) is shown. This study evaluated the stabilization error and rate of change of charge/discharge currents, the switching time from the charge mode to the discharge mode and vice versa for a single BRC and parallel BRCs. The possibility of increasing the maximum battery testing current due to the parallel connection of multiple BRCs without using a current alignment device between channels was discussed.

1. Introduction

In electric power systems (EPS) of the satellites have been widely used rechargeable batteries (RB) based on lithium-ion technology. Battery resource determines the duration of the spacecraft as a whole. On development stage the manufacturer typically perform electrical, thermal and other kinds of battery tests to estimate battery resource. Battery resource tests [1, 2] are performed under laboratory conditions through reproduction of battery charge/discharge modes and through maintaining the battery temperature at the values corresponding to operating modes. In addition, an indirect estimation of battery resource is possible by changing the output impedance of the battery.

These tests allow to estimate the following resource characteristics of the battery:

- number of charge/discharge cycles;
- battery capacity degradation dynamics during operation;
- variation of the output impedance during operation.

Quicker resource tests can significantly speed up and reduce the cost of designing and testing of the spacecraft power system. To reduce testing time, the specific battery testing methods were developed. [3, 4] These methods are based on the idea of stress tests. Reducing of the testing time is achieved by increasing the charge/discharge currents up to the maximum battery currents, including constant power discharge modes.

* This study was supported by the Ministry of Education and Science of the Russian Federation (Government Contract 14.577.21.0082, unique identifier RFMEFI57714X0082).

As equipment for battery resource tests are often used the elements of spacecraft PSS such as battery charge/discharge unit (CDU) [5] or their simulators. The disadvantages of this method include:

- charge/discharge currents and discharge powers for testing is limited by standard operation modes of systems;
- the duration of the test is comparable with the normal battery operating time;
- only a fixed number of series-connected batteries from battery pack can be tested which does not allow to test particular advanced batteries;
- all batteries in tested pack have common charge/discharge current so it is not possible to use a single battery discharge mode by constant power and to provide negative voltages on the battery.

2. The structure and operation principles of the charge-discharge test unit

To overcome the above drawbacks, the authors developed a charge-discharge test unit which allows to perform electrical, thermal and resource tests of twelve lithium-ion batteries simultaneously.

The block diagram of the proposed system is shown at figure 1. Uninterruptible power supply (UPS) powers the primary energy source (PES) and prevents unscheduled completion of the tests in the case of emergency AC network shutdown. Charge and discharge modes of the battery provided with the appropriate switching of it in the electrical circuit formed by controlled voltage stabilizer (CVS) and a switched-mode voltage converter (SVC) using switches S1-S4. Application of the CVS allows to ensure the normal operation of the SVC because the battery has a relatively low voltage and must be tested in the reverse polarity mode. Excess energy, consumed by SVC, dissipated in the load unit (LU).

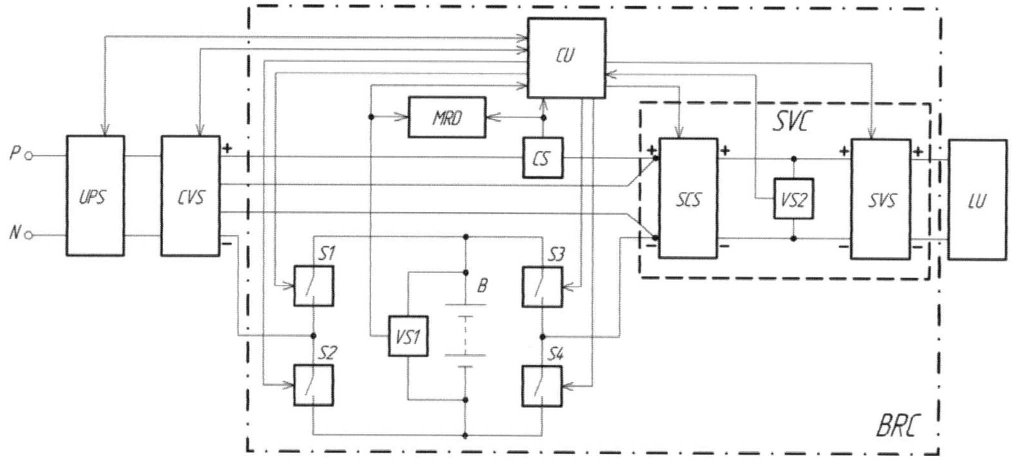

Figure 1. The block diagram of the charge-discharge test unit for the battery resource tests

Measuring and registering device (MRD) registers changes in current, voltage and temperature of the battery in time (battery temperature stabilization and measurement system are not shown in block diagram).

Necessity of battery current regulation in the wide range (0-160 A), the need for electrical isolation between SVC and LU and complexity of power high-frequency transformer manufacturing, designed for maximum battery current, made it necessary to use two stabilizers in the SVC: switched-mode stabilizer of input current (SCS) and switched-mode stabilizer of SCS input voltage (SVS). SCS consist of switched-mode current regulator (SCR), which is based on boost converter, current sensor (CS) and CU. SVS consist of switched-mode voltage regulator (SVR), which is based on full-bridge buck converter, voltage sensor (VS2) and control system [6].

A control unit (CU), in accordance with predetermined patterns and the measured values of battery voltage and current, control switches S1-S4, generates control signals for SVC, set reference for PES, and provides safe completion of tests if charge of the UPS batteries is low and if voltage in AC network is absent. CU includes a device which protects battery against charge/discharge currents and battery voltages that exceeds their maximum permissible levels. One battery research channel (BRC) consists of four switches S1-S4, voltage sensor (VS1), MRD, CU and SVC.

To provide high current at low battery voltage is necessary to maintain the SVS input voltage at the relatively high level. To this end, the stabilization of the SCS input voltage is provided by CVS, for which the feedback signal is formed by SCS input voltage. Therefore, regardless of battery voltage and the operating mode (charging or discharging of the battery) CVS stabilizes the SCS input voltage at the level of 4 V. In addition, the stabilization of the SCS input voltage enables to discharge the battery until changing the polarity.

In the boost converter voltage conversion ratio varies over a wide range depending on the current and the load impedance. [7, 8] To maintain the relatively high value of the SCS voltage conversion ratio a SVS is introduced in circuit. This SVS stabilizes the SCS output voltage.

In the charge mode (fig. 2) CU closes switches S2 and S3 while the battery and the CVS are switched to opposite-series connection. In this mode CU performs the following functions:

- setting of the SCS control law: the stabilization of the charging current or stabilization of the charging power;
- providing the stabilization of the SCS output voltage;
- setting of the CVS parameters.

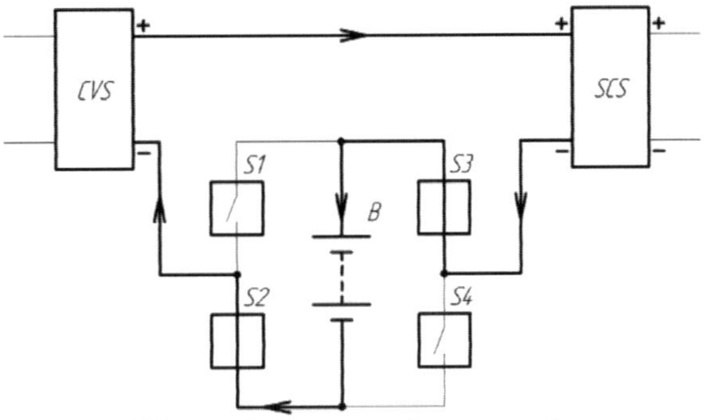

Figure 2. Battery charge mode

In the discharge mode (fig. 3) CU closes switches S1 and S4 while the battery and the CVS are switched to agree-series connection. CU performs functions similar to those described above, but corresponding to the discharge mode.

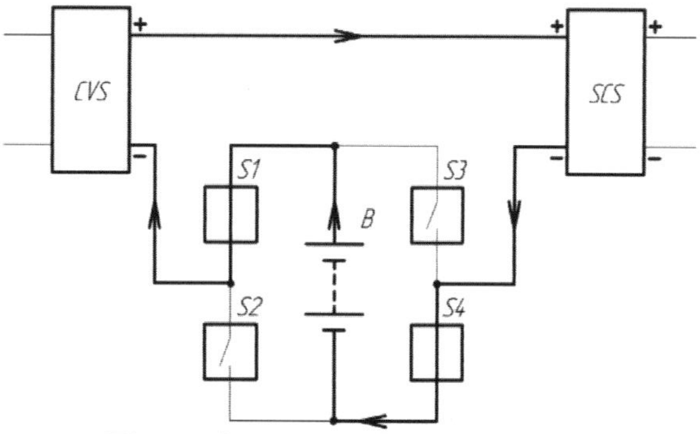

Figure 3. Battery discharge mode

The unit is implemented for parallel operation of the two BRC in the discharging mode. In this mode, the power outputs of both modules are connected to the battery and the provided discharge current range is 2...320 A. There is also the independent BRC connection mode in which provided discharge current range is 1...160 A.

An implementation of the stand is shown at figure 4. It has twelve BRC modules and the separate rack which includes CU, LU and test load for the unit testing. CU is implemented on the basis of the measuring platform National instruments NI PXIe-1082. As UPS were used EATON 9130.

Figure 4. The proposed units with CD, CCMD, SLU and test load for the stand testing

With the development of technology and the increasing of total capacity of the batteries, the maximum allowable battery charge/discharge currents are also increasing. In this connection there is a need for charge/discharge devices with the maximum currents above hundred amperes. Time intervals of the discharge and charge cycles under stressful conditions may be relatively small and may be about a few seconds. Therefore, performance (delay) of the converters which are used in load units must be sufficiently large (small) that does not significantly affect the magnitude of the calculated capacity or battery power. Measurement errors accumulated over several months of the resource tests and may affect the accuracy of the resource tests that requires greater accuracy.

The stand is designed with consideration of these comments and each BRC has the following specifications:

- battery voltage variation – from −5 V to +5 V (minimal step – 0.05 V);
- battery charge/discharge current variation range from 1 A to 80 A (minimal step – 0.1 A);
- battery discharge current variation range from 1 A to 160 A (minimal step – 1 A);
- time interval variation – from 1 s to several hours in increments of no more than 1 s, the interval representation error within ± 0.6 %.

CU in combination with the battery current and thermal sensors processes the errors:

- charge/discharge current recording error – within ±0.2 %;
- battery voltage recording error– within ±5 mV;
- time interval recording error – within ±0.1 s (in accordance with the duration of the intervals in the program).

3. Experimental results

During the test stand performs the program recorded in the CU. The program is executed step by step. The desired values of stabilized current or power and the duration of the step are set at each step. Thus, it is necessary to evaluate for compliance with the technical requirements:

- charge/discharge current stabilization error at each step in the prescribed current range for single BRC;
- the rate of change of the charge/discharge current for a single BRC;
- the total charge/discharge current stabilization error in parallel BRC operation;
- current mismatch between the modules in parallel operation of BRC;
- switching time from battery charge to discharge mode, and vice versa, in parallel BRC operation.

To measure the charge/discharge current stabilization error the short-circuit tests were performed on single BRC. The duration of the program step – 5 s, a discrete battery current level step - 0.5 A, current range - 0...160 A. After starting the program, the CU recorded the predetermined current and the voltage value from the sensor at each step with the period of 0.1 s.

Reduced charge/discharge current stabilization error of the single BRC (figure 5) calculated as:

$$\gamma_1 = (I_p - U_m / R_m \cdot k) / I_{max1} \cdot 100\%, \quad (1)$$

where I_p – predetermined discharge current, U_m - current sensor voltage measured on the resistance $R_m = 12.5$ Ohm; $k = 1000$ – current sensor convert ratio, $I_{max1} = 160$ A – the maximum current of the single channel.

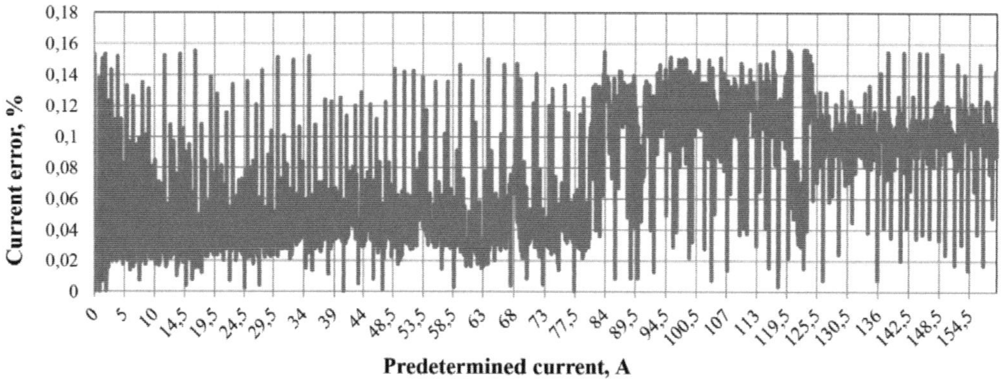

Figure 5. Reduced charge/discharge current stabilization error of the single BRC

The short-term deviation of the BRC current is observed when the discharge current changes discretely which leads to additional lag and overshoot. This deviation leads to short-term increase in the current error. However, resulted reduced charge/discharge current stabilization error of the single BRC over the entire range of the current does not exceed 0.16 %.

To measure the accuracy of the BRC in parallel mode tests were performed on a similar program: the step duration is the same but the range of current change was 70-320 A. In addition, each BRC had stabilized the current in a half of the predetermined value I_p. During the test with period of 0.1 s the voltage was recorded by the first BRC current sensor (U_{m1}) and by the second BRC current sensor (U_{m2}).

For each measured value (figure 6) was calculated the total reduced current stabilization error γ_2:

$$\gamma_2 = (I_p - (U_{m1} - U_{m2}) / R_m \cdot k) / I_{max2} \cdot 100\%, \qquad (2)$$

where $I_{max2} = 320$ A – the maximum discharge current.

In addition, the mismatch between discharge currents δ of parallel BRC was calculated:

$$\delta = (U_{m1} - U_{m2}) / R_m \cdot k / I_{max2} \cdot 100\%. \qquad (3)$$

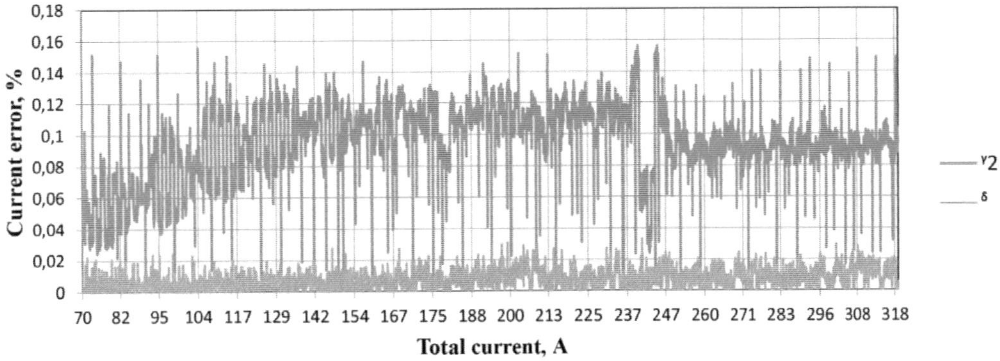

Figure 6. The total reduced current stabilization error and the mismatch between discharge currents

The total reduced current stabilization error γ2 of parallel BRC in the second range does not exceed 0.16 %. The mismatch between discharge currents δ of parallel BRC in the second range does not exceed 0.03 %.

Since the total reduced current stabilization error and the mismatch between discharge currents in both ranges are small, in parallel operation of BRC (as well as for a single channel) is provided high precision of stabilization of the total discharge current and good current equalizing between the two BRC.

Since the currents of several parallel BRC are nearly equal, the overload is eliminated in parallel BRC as well as the need to use a special device to equalize the currents between the BRC.

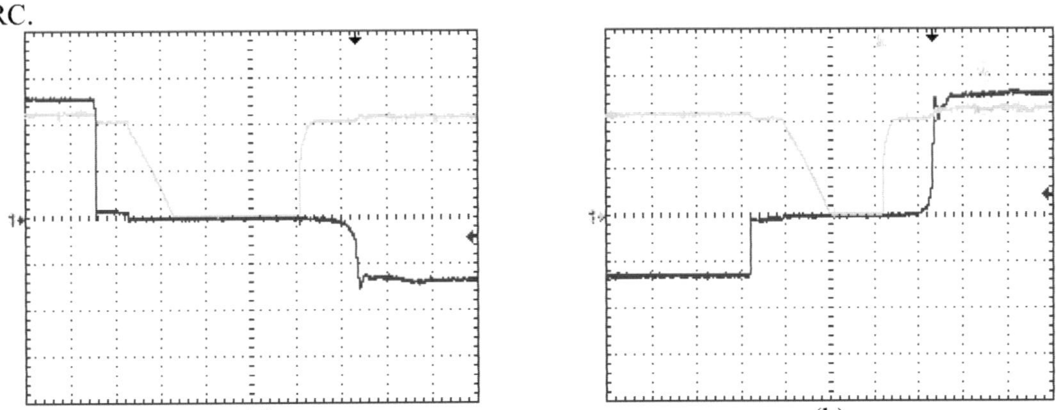

(a) (b)

Figure 7. Switching (a) from the charging mode (50A) to discharging mode (25A), (b) from the discharging mode (25A) to charging mode (50A) (1 (blue) – BRC current, 2 (green) – SVC voltage),
scale: x-axis – 100 ms/div, y-axis – 20 A/div

To evaluate the switching time between the charging and discharging modes for the single BRC the short-circuit test have been performed. The program switched charging mode to discharge mode at charge current of 50 A, then - vice versa, at discharge current of 25 A (figure 7a), the program switched discharge mode to charge mode at discharge current of 50 A, then - vice versa, at charging current of 25 A (figure 7b).

The switching time from the charging mode to discharging, and vice versa, is the sum of the BRC turn-off time, delay time and turn-on time and does not exceed 0.6 seconds. The BRC turns off fast enough but turn on of the BRC is delayed since the CU sends a command to switch the channel only when the SVC voltage is reached a desired level. The delay time is the major part of the switching time and is determined by inertial properties of the SVC.

To evaluate the switching time between the charging and discharging modes for the parallel BRC operation the short-circuit test have been performed. The program switched charging mode to discharge mode at charge current of 50A, then - vice versa, at discharge current of 50A (figure 8a), the program switched discharge mode to charge mode at discharge current of 50A, then - vice versa, at charging current of 50A (figure 8b).

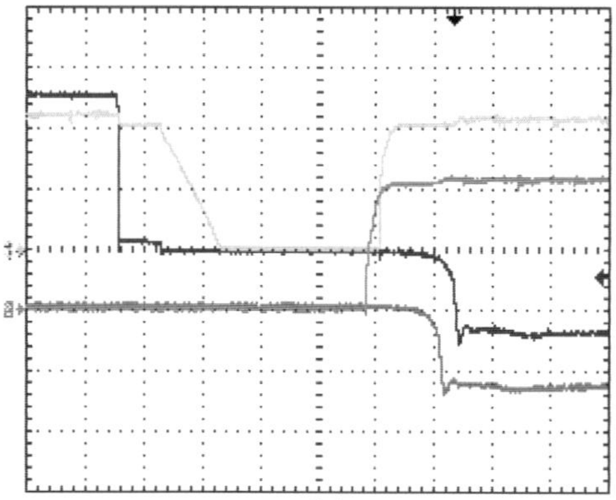

(a)

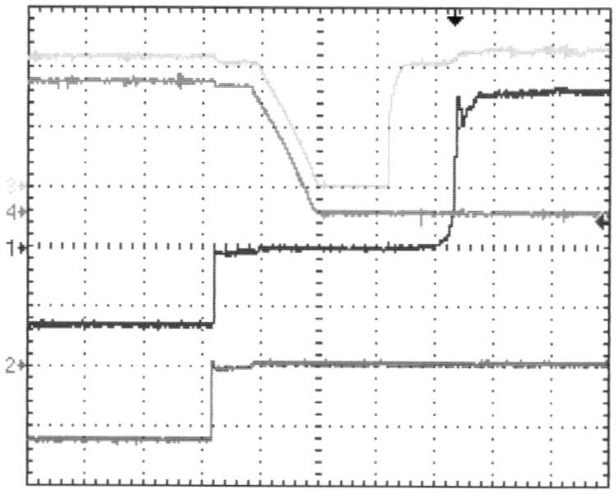

(b)

Figure 8. Switching (a) from the discharging mode (50A) to the charging mode (50A) during the parallel operation of the two BRCs, (б) from the charging mode (50A) to the discharging mode (50A) during the parallel operation of the two BRCs (1 (blue) – BRC 1 current, 2 (green) – BRC 2 current, 3 (red) – SVC voltage of BRC 1, 4 (magenta) – SVC voltage of BRC 2),
scale: x-axis – 100 ms/div, y-axis – 20 A/div

As in the previous case, the switching time from the charging mode to discharging, and vice versa, is the sum of the BRC turn-off time, delay time and turn-on time and does not exceed 0.6 seconds. Time and rate of switching are largely depending on the inertial properties of the SVC. The delay time is most of the switching time and can be reduced by using a high-speed SVCs and by enhancement of SVC control methods.

4. Conclusion

In this study the charge-discharge test unit was described. This unit allows to perform electrical, resource and thermal testing of several lithium-ion batteries simultaneously. The achieved accuracy of the unit is relatively small and corresponds the requirements for the resource tests. The switching time from charging mode to discharging, and vice versa, depends on the delays introduced by the power source. Excluding these delays will reduce the switching time to 0.2-0.3s. Parameters of parallel operation of the modules confirm the prospect of increasing the maximum testing current for even larger values of the total battery charge than currently in use.

References

[1] Korotkih V V, Kudryashov V S, Nesterishin M V, Pochebut D V, Shanavrin V S 2010 *Method of performing service life tests of space storage batteries and device to this end* Patent RU 2390477

[2] Palanisamy T G, Singh H N, Atehortua H M, Hoenig S 2007 *Method and apparatus for determining the available energy of a lithium ion battery* Patent US 7248022

[3] Groot J 2012 *State-of-Health Estimation of Li-ion Batteries: Cycle Life Test Methods* (Goteborg: Chalmers University of Technology Press)

[4] 2001 *PNGV Battery Test Manual* (Idaho: Idaho National Engineering and Environmental Laboratory)

[5] Guiheen J V, Singh H, Palanisamy T G 2004 *Method and apparatus for determining the state of charge of a lithium-ion battery* Patent US 6774636

[6] Hatsuyado H, Hoshi N 2014 *A high boost ratio DC-DC converter for low voltage fuel cell* (IEEE International Conference on in Renewable Energy Research and Application (ICRERA))

[7] Wen J 2006 *A new interleaved isolated boost converter for high power applications* (Twenty-First Annual IEEE in Applied Power Electronics Conference and Exposition (APEC '06))

[8] Asfani D A 2014 *Experimental design boost ratio converter with hybrid transformer for hybrid power source charging station* (IEEE International Conference on in Electrical Engineering and Computer Science (ICEECS))

Building automation system of payment platform weight component for large spacecraft reflector

I V Kovalev[1], J O Badanina[2]
[1] Siberian State Aerospace University named after Academician MF Reshetnev Russian Federation, 660037, Krasnoyarsk, pr. Them. gas. "Krasnoyarsk worker", 31
[2] Siberian State Technological University, Russian Federation, 660049, Krasnoyarsk, pr.Mira 82

E-mail: kovalev.fsu@mail.ru, secret398@mail.ru

Abstract. Considered Design and the logic of opening large convertible antenna. The necessity of compensation weight component in the assembly and testing of the design. Given the logic of the movement elements of power spokes, concluded that the use of the tracking system to compensate for the weight component.
The analysis of the existing equipment and control systems. Produced selection of the manufacturer of automated equipment that meets the stated objectives of management and control. It is concluded that the design component of the weight compensation system based on servo controllers and sensors combined platform automation, controlled by special software.
The structure of the platform automation, consistent workflow testing. It defines the principles of interaction between subsystems of the weight compensation component for receiving, processing and monitoring of process parameters testing. It is concluded that the proposed system can be integrated into the automation system and the perspective of process control testing of disclosure of large spacecraft.

Keywords: system of compensation weight component; spacecraft, bulky reflector, the control system, the model, the automation platform module, servo, controllerб, software.

Introduction. In connection with the introduction of the industry bulky mechanical systems spacecraft became necessary to find new, but it is quite simple and effective ways of working out these mechanical systems at the stage of terrestrial experimental testing, since the use of traditional methods either do not meet the requirements, or cannot be used due to the constructional features.

At various spacecraft conducted a series of experiments for the deployment of large-sized antenna structures. Analysis of existing in the world market of large transformable reflectors showed that, despite some progress in this area, only a few of them are able to ensure the accuracy of weightlessness respective predetermined shapes and the necessary rigidity. It is not possible always to achieve a reliable disclosure. The reason is the violation of technology of the following steps:

- manufacture of parts and assemblies based on the resource characteristics of the structure;
- high-precision assembly bench on the complex;
- system the test antenna.

For all of the existing spacecraft at the present time must be performed a number of tests, which require the elimination of the weight component for constituent elements of a design, to avoid deformation or destruction under the influence of gravity. To fulfill this requirement use system of compensation weight component. These systems may find their application in the test operation of mechanical systems at disclosure processes transformed antennas, truss reflectors in different shapes and sizes, of solar panels and other designs. [1-4]

Appointment. The compensation systems for weight component intended to compensate for the moments of the forces acting with respect to the root of the hinge assembly relative to the intermediate joint assembly relative to the hinge assembly canting spokes by weight of units, and the moving parts of the reflector, including separately from the forming structure of tapes deployment mechanism from canvas and other essential structural components of the reflector. Configure and verify the operation of the reflector.[5]

The system of compensation weight component should provide a job with a reflector, a key element of the support structure, which is the power spokes (Figure 1), which define the desired stability of the work surface under the influence of factors operating. Consider, the basic principles of the compensation system of modules required to provide compensation for the weight of the constituent elements of the antenna power spokes.

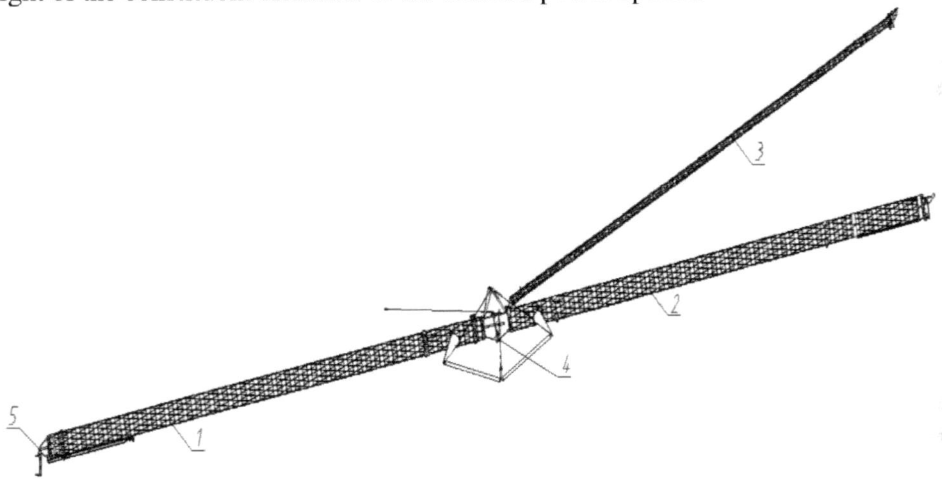

Figure 1. Power spoke
1 - root element; 2 - terminal units; 3 - folding unit;
4 - Intermediate hinge assembly; 5 - root hinge assembly.

To ensure the required dimensions in the transport position, the power spike has a convertible structure and is driven to its working position by means of a sliding mechanism of the mast. To construct the trajectory of the center of mass at the time of disclosure of the spokes, a table 1 to 20 units, and the provisions of the strut relative to each other during the disclosure of the antenna.

Table 1 - The theoretical parameters of kinematics motion spokes and change the position of the center of mass during deployment spokes

Conditional status spokes links	Angles disclosure		Spoke root		Spoke end		Struts	
	α1	α2	Y, mm	X, mm	Y, mm	X, mm	Y, mm	X, mm
1	0	0	2,48	2313,37	167,1	2330,16	271,39	1534,75
2	0	10,247	2,48	2313,37	465,55	2371,19	709,66	1607,02
3	9,024	10,247	365,31	2284,35	831,72	2268,82	952,95	1475,81

4	15,143	20	606,71	2232,4	1355,26	2180,68	1526,52	1396,96
5	20,827	30	824,85	2161,32	1856,49	2070,93	2086,24	1302,31
6	26,143	40	1021,52	2075,62	2328,41	1948,97	2620,16	1201,69
7	31,043	50	1195,08	1980,78	2768,04	1827,27	3125,07	1108,88
8	35,301	60	1338,88	1886,56	3167,92	1728,72	3595,02	1049,65
9	38,624	70	1445,95	1805,8	3519,72	1685,15	4022,89	1060,35
10	41	80	1519,57	1744,3	3822,27	1713,52	4403,89	1161
11	42,875	90	1575,84	1693,64	4084,82	1794,79	4738,68	1330,01
12	44,751	100	1630,45	1641,13	4318,52	1894,75	5031,51	1527,03
13	46,902	110	1690,88	1578,8	4529,99	1988,05	5286,52	1721,15
14	49,401	120	1758,12	1503,57	4721,07	2061,81	5505,96	1895,95
15	52,239	130	1830,41	1414,67	4891,15	2110,11	5690,6	2043,4
16	55,518	140	1908,33	1307,66	5043,81	2118,17	5845,57	2145,91
17	59,556	150	1995,67	1170,04	5189,93	2051,64	5886,04	2797,27
18	64,682	160	2092,22	987,06	5336,97	1873,35	5967,21	2675,42
19	71,247	170	2191,29	741,58	5482,94	1543,53	6064,13	2381,61
20	79,906	180	2278	403	5613	991,2	6174,5	1842,6

$\alpha 1$ - the angle between the axis of the mast and the needle root; $\alpha 2$ - the angle between the needle and the needle root end.

According to the data of Table 1 plotted the center of mass of the root needles, needles and strut end (Graph 1).

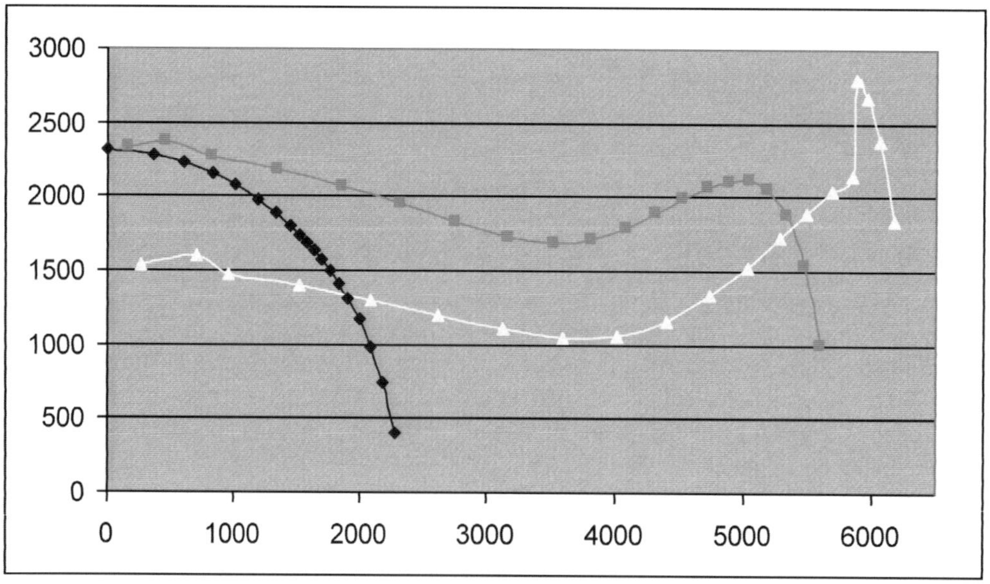

Graph 1. The center of mass of the root spoke, spoke and end strut.

The graph shows that the trajectory of the links and the strut rather complicated, requiring the use of the tracking system to compensate for the weight of the component elements in the process of working out the disclosure antenna in ground conditions.[6-8,10]

Production system management tasks. To implement compensation component of the spokes during assembly and testing, it is necessary to organize a three-point connection, are in the theoretical center of mass of the root, and end folding units with the module component of the weight compensation system. Communication with the executive system compensation follower weight component will be implemented through a flexible inextensible sling.

Actuators compensation system of the weight component will be three carriages moving along the rail and ensure a certain predetermined deviation from vertical flexible lines, as well as three drive winder flexible lines to changes in the length of the flexible lines, changing the coordinates of the centers of mass units. From the center of gravity of links and their trajectories are added acceptable compensation system moves carriages weight component.

The module compensation system of the weight component must meet the following basic conditions: to ensure the maintenance of a predetermined force in flexible slings and control, tracking deviation from vertical flexible slings and control the angle of deflection on the entire trajectory of the centers of mass units. The reaction of the actuator system must provide the necessary dynamics in the process of opening an antenna.[4,9]

The proposed platform automation. Analysis of domestic and foreign market of automated equipment showed that domestic producers lack the necessary range.

The equipment satisfies the necessary conditions only firms "Siemens", "Rockwell Automation" and "Mitsubishi Electric".

Analysis of equipment from Siemens shows that it does not satisfy the mass-dimensional parameters and the methods of construction of the SCADA-system.

Consider the equipment of firm "Rockwell Automationr". Distributors of this equipment are ready to put it, but does not deal with technical support and maintenance projects.

Analysis of the equipment of the Japanese firm Mitsubishi Electric demonstrated that it meets the requirements for dimensional parameters, specifications and software products that can implement an appropriate management system and program shell for the operator system compensation.

In order to implement compensation weight component is offered on the basis of an automation platform MELSEC System Q Mitsubishi Electric. Mitsubishi System Q platform is based on advanced multiprocessing technologies, one system will have multiple processors - processor PLC, motion control processor, the processor control program based on SCADA-system with the possibility of full integration of requests for management and exchange of data on a single platform. In addition a high degree of system availability by using redundant master devices in communication networks, as well as recognition of the break lines and the ability to "hot swap" that allows module replacement during service system.

The traffic management controls synchronizes the connected servo amplifiers, servo motors and a set of required sensors are integrated into the control system for the continuous visualization and tracking the positions of the centers of mass units spokes.

Block diagram of automation presented in Figure 2.

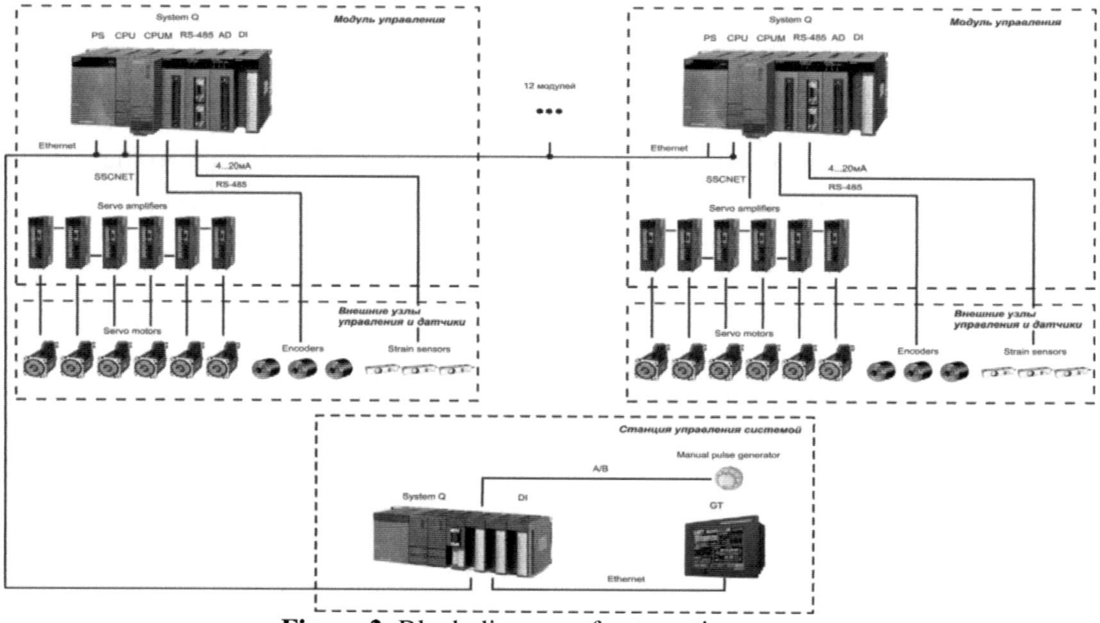

Figure 2. Block diagram of automation.

Each module includes a separate control unit, consisting of PLC, motion control module, analogue input module, the module of digital inputs, outputs module, power supply and servo amplifiers located on the chassis. All modules are connected to the base station control with Ethernet. Basic control station consists of a PLC module MES-interface switching Ethernet module and a personal computer with the operating system for positioning controllers and SCADA system. SCADA-system is implemented on the MX4 SCADA - a fully functional system of monitoring and data collection, based on the interface of Windows, which considerably shortens the learning curve, enabling the user to quickly adapt and respond to changes in the process. The SCADA-system programmed basic functions, including locking, alarm, archive data and reports, provides a simple, but important solutions.[12,13]

Visualization works with the control program executed on the operator panel Mitsubishi Electrical GOT2000, which provides the human-machine interface that provides access to process data, allows you to track, manage, and as necessary to change the parameters of the compensation weight component. Automatic control of the process of compensation weight component is constructed in such a way that on the computer screen displays the main menu system, containing functional groups:

- Mode of assembly;
- Test mode;
- Statistics;
- Configuration and diagnostics;
- Active alarms;
- Archive.

Automatic compensation control weight component is to control and maintain the desired process parameters, by the system control operation of the actuators.

With regular tests designed system runs according to the specified algorithm, maintaining the value of certain efforts and the vertical position of the flexible connection of the entire trajectory of the disclosure of the spokes of the antenna, and also takes into account the impact of dynamic jerks and prescribed emergency situation.

Management system is designed so that the control program set the basic parameters of the compensation weight is:

- The value of the desired tension of the flexible lines;
- Needed speed (or the laws of change of speed) servo tracking upright flexible straps.

Servo motors allow you to set any of the known laws of regulation and set any parameters for a smooth acceleration - deceleration, working out signals from sensors measuring the angle and load cells.

Conclusion. As a result of this article, it concluded that the proposed system can be integrated into the automation system and in the perspective of process control testing of disclosure of large spacecraft antennas.

The structure of the compensation system automation platform weight component, consistent process of testing. Designed intuitive man-machine interface, with the possibility of remote monitoring and control settings.

Using the selected automation platform will solve the task management of the compensation system, which will increase its reliability, expand functionality, will be more convenient to use, increase productivity and reduce errors operator to reach the desired level of efficiency of the system as a whole. Such an assessment can confidently assert the qualitative improvement of the system of compensation weight component, which will have greater flexibility and dependability

References:

1. Gulyaev VI, Gaydaychuk, Cherniavsky AG, Lenin L. On the dynamics of the unfolding of large-sized reflector / Applied Mechanics. - Kiev, 2003 -39. 9 -S.109-115.

2. John M. Hedzhepet. Design for precise remote antennas disclosed. Final Report. Prepared by the National Aerospace Research Center and Langley contract № NASA-18567. 1989.

3. Kovalev IV, Kikot JO Mobile system simulate weightlessness for large spacecraft. // Herald Siberian State Aerospace University named after Academician MF Reshetnev, Issue 4 (56). 2014. Pp 173-178.

4. Patent №233970 G61M19 C2 / 00 V64P7 / 00. Drozdov AA Agashkin SV Mikhnev MM, Ushakov, AR The device simulating weightlessness with Flexible design elements. RU, 27.09.2008.

5. Goldobin NN Methodology to evaluate the shape of the reflecting surface of bulky radio convertible reflector spacecraft // Herald SibSAU. 2013. Vol. 1 (47). Pp 106-111.

6. Debda DE, GY Pyatibratov Problems of creation of the combined compensation systems gravity objects obezveshivaniya // Novocherkassk: SRSTU, 2000 - 32 E.-Dep. VINITI 16.02.2000, № 396-V00.

7. Poduraev Yu Mechatronics: fundamentals, methods, applications. // M .: Engineering, 2007. 256 pp.

8. R. Dorf, Bishop R. Modern control systems / per. from English. BI Kopylov. Moscow: Laboratory of Basic Knowledge, 2002. 832 pp.

9. Romanov AV, NA Testoedov Basics of designing information control and mechanical systems of spacecraft. Textbook. Edited by Doctor of Technical Sciences, Professor VD Atamasova // LA ANO "Professional", 2015. - p.240.

10. Patent №1828261 A1 G01M19 SU / 00.Anuprienko GE, Karpachev YA, Kuhotsky LM Mishneev AA Pavlyuk VN Rudykh Yu, Y. Savenko N. The device obezveshivaniya elements. RU, 20.09.1996.

11. Patent №5848899A. Ian P. Howard. Method and device for simulating weightlessness.US, 15.12.1998.

12. Platform Mitsubishi Electric MELSEC System Q and its components [Electronic resource]. URL: http // mitsubishielecric.ru

13. Software GX IEC Developer and MX OPC Server [electronic resource]. URL: http // mitsubishielecric.com

Model prototype utilization in the analysis of fault tolerant control and data processing systems

I V Kovalev[1], R Yu Tsarev[2], D V Gruzenkin[2], A V Prokopenko[2], A N Knyazkov[2], V D Laptenok[1]
[1] Siberian State Aerospace University named after Academician M.F.Reshetnev
31 "Krasnoyarskiy Rabochiy" prospect, Krasnoyarsk, 660037, Russia
[2] Siberian Federal University, Russia
79, Svobodny Prospect, Krasnoyarsk, Russia

E-mail: tsarev.sfu@mail.ru

Abstract. The procedure assessing the profit of control and data processing system implementation is presented in the paper. The reasonability of model prototype creation and analysis results from the implementing of the approach of fault tolerance provision through the inclusion of structural and software assessment redundancy. The developed procedure allows finding the best ratio between the development cost and the analysis of model prototype and earnings from the results of this utilization and information produced. The suggested approach has been illustrated by the model example of profit assessment and analysis of control and data processing system.

1. Introduction

Inclusion of structural and software redundancy is one of the primary approaches to the increasing of the fault tolerance of control and data processing systems [1, 2]. However, in this case the cost exposure increases, due to the fact that frequently it is not possible to get complete information during the systems formation of the class.

The problem is that there are two sources of uncertainty. The first one is expressed by the probability of the fact that research with using model prototype leads to the selection of system implementation variant on condition that this variant will be unsuccessful. The second one is expressed by the probability that model prototype utilization leads to the variant selection on condition that it will be successful [3-5].

One of the main methods of the profit expected value calculation with the use of model prototype during the choice of preferred data control and processing system implementation is utilization of Bayes' theorem and its special cases [6, 7]. In addition, real costs under the different expenses for modeling and system debugging, which provide the various reliability levels for fault tolerance prediction of their individual elements, can be evaluated [8, 9].

Model prototype realization enables to obtain the information, which is sufficient for detection of the volume of input structural and software redundancy. Due to this fact, the alternative, which maximizes the profit of control and data processing system implementation, can be recommended.

2. Methodology

In the general case, obtaining of complete information of the real situation by prototype development method or other knowledge increasing methods is not possible.

There are two sources of uncertainty, which can be described through probability:

- $P(RM|UC)$ – the probability of that the research (R) leads to the choosing of radical (M) method providing that in real situation this method will be unsuccessful (UC – unsuccessful case);
- $P(RM|SC)$ – the probability of that the research (R) leads to the choosing of radical method providing that in real situation this method will be successful (SC – successful case).

As the prototype can not always adequately reproduce some of the technical details or can be just an approximate model of the real system, generally the probability $P(RM|UC)$ is different from zero.

Thus, the prototype can confirm choosing of radical alternative although in the reality the result can be unsuccessful.

The probability $P(RM|UC)$ is usually not equal to zero, as the prototype can contain errors which will be bug fixed in the real control and data processing system.

It indicates that there is a probability that the prototype will show redundancy invalidity and therefore provide support for choosing of conservative alternative whilst the radical one may become successful.

Based on the values of the variables according to probability theory we can get the following ratios:

$$P(RM) = P(RM|SC) \cdot P(SC) + P(RM|UC) \cdot P(UC); \tag{1}$$

$$P(RC) = 1 - P(RM); \tag{2}$$

$$P(SC|RM) = P(RM|SC) \cdot P(SC) \,/\, P(RM); \tag{3}$$

$$P(UC|RM) = 1 - P(SC|RM)$$

Formula (1) corresponds to the case in which two variants of radical alternative can be chosen after prototype research. These variants are choosing of radical method when it is successful (the probability of this event is equals to $P(RM|\, SC) \cdot P(SC)$), and choosing of the radical method when in actual truth it leads to failure (the probability is equals to $P(RM|UC) \cdot P(UC)$.

Formula (1) equates probability of choosing radical method to the sum of probabilities of two mutually exclusive cases.

Formula (2) adds to (1): it corresponds to the prototype research, which leads to choosing of conservative alternative (RC).

Formula (3) shows the probability of a successful case under the condition of radical method and that in the result the prototype research leads to its choice and is determined in accordance with $P(SC|RM) = P$(choosing of RM – radical method, in case of its successful completion) / P(choosing of RM).

Formula (3) is a special case of Bayes' formula and the main formula for determination of profit expected value with the utilization of imperfect prototype for the choosing of preferred development method of data control and processing system.

2.1. Formalization of procedure of full information profit expected value determination

Let us assume that m is a set of alternatives (implementation variants of control and data processing system). A_1, A_2, …, A_m in the situation which has n possible conditions S_1, S_2, …, S_n, with probabilities values $P(S_1)$, $P(S_2)$, …, $P(S_n)$, profit value of alternative A_i choosing in condition S_j are defined by gain matrix V.

The goal is to choose an alternative with the maximal expected value of gain. According to this definition, the profit can be expressed as dollars, quality indicator units, and utility function.

Based on expert based estimation of each condition probabilities profit expected value (*PEV*) when alternative A_i is chosen:

$$PEV(A_i) = P(S_1)V_{i1} + P(S_2)V_{i2} + \ldots + P(S_n)V_{in}$$

Let us choose the alternative of maximum expected value:

$$PEV_{no\ info} = \max\ [P(S_1)\ V_{i1} + \ldots + P(S_n)\ V_{in}],\ i = 1, \ldots, m.$$

Let us choose the alternative with maximum profit when having full information on each case:

$$PEV_{full\ info} = P(S_1)\ (\max\ V_{i1}) + \ldots + P(S_n)\ (\max\ V_{in}),\ i = 1, \ldots, m.$$

Then expected value of profit on full information (*FI*):

$$PEV = PEV_{full\,info} - PEV_{no\,info} = \sum_{j=1}^{n} P(S_j)(\max V_{ij}) - \max \sum_{j=1}^{n} P(S_j)V_{ij}, i = 1,\ldots,m.$$

2.2. Formalization of procedure of incomplete information profit expected value determination

An alternative, which maximizes the profit, can be always recommended when the research provides the full information for task of choosing. RA_i alternatives, recommended in the result of research, correlate with conditions S_j by the following ratios:

- $P(RA_i|\ S_j) = 1$, if A_i maximizes the profit for condition S_j,
- $P(RA_i|\ S_j) = 0$, in the contrary case.

Recommendation of RA_i alternative is based on incomplete information on the conditions. Thus, for each alternative A_i and condition S_j the probability $P(RA_i|\ S_j)$ reflects the degree of possible departure from the ideal case (0 or 1) for the appropriate combination A_i and S_j. The sum of probabilities $P(RA_i|\ S_j)$ on each conditions S_j must equal to 1.

The general formula for profit expected value calculation in the task solution of alternatives choosing among $RA_1, RA_2, \ldots, RA_m$ is:

$$PEV(RA_1, RA_2, \ldots, RA_m) = \sum_{i=1}^{m} P(RA_i) \cdot PEV\,(\text{profit on acceptance of } RA_i) =$$

$$= \sum_{i=1}^{m} P(RA_i)\left[\sum_{i=1}^{m} P(S_j \mid RA_i)V_{ij}\right]. \quad (4)$$

For calculation of $PEV(RA_1, RA_2, \ldots, RA_m)$, where RA_i are RM and RC, S_j – SC and UC, the general formulas for the calculation of $P(RA_i)$ and $P(S_j|RA_i)$ according to the known values $P(S_j)$ and $P(RA_i|S_j)$ are needed. These formulas are the following:

$$P(RA_i) = \sum_{j=1}^{n} P(RA_i \mid S_j)P(S_j), \quad (5)$$

$$P(S_j) = P(RA_i \mid S_j)\, P(S_j) \,/\, P(RA_i). \quad (6)$$

The ratio (6) is a standard form of Bayes' formula.

3. The procedure

The following procedure is suggested for the definition of the best ratio between the expenses on research making and profit on the application results of the information produced.

Step 1. Defining the set of alternatives of management information system development methods $A_1, A_2, \ldots, A_m$.

Step 2. Determining of all possible situations $S_1, S_2, \ldots, S_n$, which may influence the result of methods application.

Step 3. Defining the V gain matrix elements, where V_{ij} – gain of method A_i utilization in the case S_j.

Step 4. Defining the probabilities $P(S_j)$ of each situation S_j.

Step 5. $PEV_{no\ info}$, $PEV_{full\ info}$ and $PEVFI$.

Step 6. If the value $PEVFI$ can be neglected, then the additional research may not be pursued. In this case, the method providing the maximum $PEV_{no\ info}$ should be chosen, then its implementation should be started.

Step 7. Definition of $P(RA_i \mid S_j)$ for each research type, i.e. the probability that it will lead to the recommendation of alternative A_i in the situation S_j.

Step 8. Calculation of PEV on utilization of information produced during the research k, $PEV(R_k)$ according to the formulas (4)-(6).

Step 9. Calculation of the real costs for each research.

4. Results and discussion

Let us consider an example of the real costs assessment of control and data processing system under the different expenses on the prototype development.

IOP Conf. Series: Materials Science and Engineering **122** (2016) 012017 doi:10.1088/1757-899X/122/1/012017

The prototype cost and the correspondent values of profit and system real costs expected values are given in Table 1. Here EV_1 means the expected value of profit, EV_2 means the expected value of profit on information, and EV_3 means the expected value of real costs under prototype development.

Table 1. Expected value of real costs and prototype cost

Prototype cost, thousand c. u.	$P(RM\|UC)$	$P(RM\|SC)$	EV_1, thousand c. u.	EV_2, thousand c. u.	EV_3, thousand c. u.
0	0,4	0,6	57	0	0
5	0,3	0,7	66,5	9,5	4,5
10	0,25	0,75	71,25	14,25	4,25
20	0,1	0,9	85,5	28,5	8,5
30	0	1	95	38	8

There is a possibility to increase research results reliability, take off some sources of uncertainty, decrease $P(RM|UC)$ up to 0.1 and increase $P(RM|SC)$ up to 0.9, when spending 20 thousand c.u. on prototype development. In this case, the expected profit equals to 85.5 thousand c.u. and expected value of profit on information produced during the prototype research equals to 28.5 thousand c.u. Real costs amounts 8.5 thousand c.u.

The dependency between prototype cost and real cost of control and data processing system is shown in Figure 1.

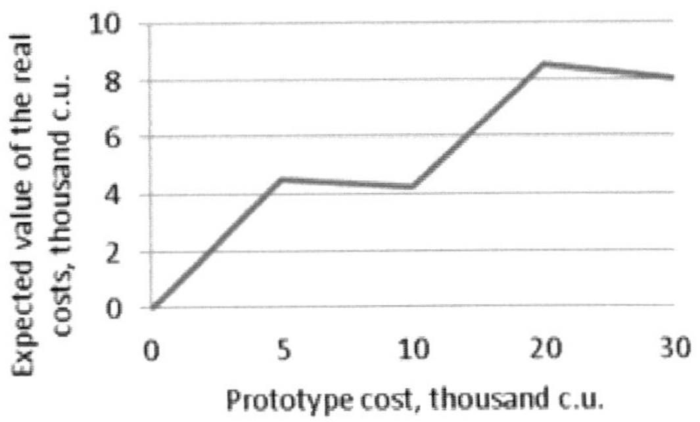

Figure 1. Dependency between real costs and prototype cost.

It may be concluded that the expenses on prototype implementation equal to 10 thousand c.u. is the best decision. Low costs do not provide reasonable profit. High costs provide more information, but at the high price, which decreases profit.

5. Conclusions

The suggested procedure of information profit determines the whole series of steps for the detection of necessary analysis of costs and income during fault tolerant control and data processing system implementation.

This procedure serves to the decision of many key problems during the fault tolerant control and data processing system composition. In particular, it allows determining the necessary tools for the following steps:

1. Research of project feasibility (users questioning, conceptual analysis, modeling, speculation, job assignment), before making the final choice of particular system variant.
2. Analysis of the alternative variants of control and data processing system realization implementation (job and tasks assignment, walkthroughs, performance analysis), before making the final choice of particular system variant.
3. Risk analysis (imitation, prototype research, user interaction study, task assignment, modeling, sensitivity test), before the concretization of control and data processing system requirements and project implementation.
4. Verification and assertion (requirement development, design, critical testing, real tests), before exploitation of the developed control and data processing system.

Model prototype development as a tool of assessment and analysis of control and data processing systems enables calculation of profit on its implementation, and the necessary detail level and required investment for system model prototype development and research.

6. References

[1] Dwyer V M 2012 Reliability of various 2-out-of-4:G redundant systems with minimal repair *IEEE T. Reliab.* **61** 170-179

[2] Golubev I M, Tsarev R Ju and Semenko T I 2005 N-version software systems design *11th International Scientific and Practical Conference of Students, Postgraduates and Young Scientists; "Modern Techniques and Technologies", MTT 2005 – Proceeding* pp. 147-149

[3] Yang Q, Hong Y, Chen Y and Shi J 2012 Failure profile analysis of complex repairable systems with multiple failure modes *IEEE T. Reliab.* **61** 180-191

[4] Mougouei D 2013 Goal-based requirement engineering for fault tolerant security-critical systems *Int. J. Soft. Eng. Appl.* **7** 1-14.

[5] Martínez-Álvarez A, Restrepo-Calle F, Vivas Tejuelo L A and Cuenca-Asensi S 2013 Fault tolerant embedded systems design by multi-objective optimization *Expert Syst. Appl.* **40** 6813-6822

[6] Ion D C and Geacar C 2012 Estimation of the cost function using Bayesian approach *UPB Sci. Bull.* D **74** 53-60

[7] Abbas A E, Hazelrigg G A and Alkindi M 2012 Bayesian inference for the demand of engineering products *Proceedings of the ASME Design Engineering Technical Conference* **7** pp. 785-792

[8] Amari S V 2013 Optimal design configurations of fault-tolerant systems *Proceedings - Annual Reliability and Maintainability Symposium*

[9] Goodall R, Dixon R and Dwyer V M 2006 Operational reliability calculations for critical systems *IFAC Proceedings Volumes (IFAC-PapersOnline)* **6** pp. 771-776.

Formation and control system by the specialized data in information networks

I V Kovalev, P V Zelenkov, M V Karaseva, V V Brezitskaya, D I Kovalev
Reshetnev Siberian State Aerospace University
31 "Krasnoyarsky Rabochy" prospect, Krasnoyarsk, 660037, Russia.

E-mail: zelenkow@rambler.ru

Abstract: The paper considers the problem of collection, processing and control by the highly specialized information. The system offers the correct solutions for a multilingual representation of information in the Internet. It is proposed to use the implementation technology of the information control systems based on multi-agent approach.

Nowadays an active development of the network technologies occurs. The most important point in these technologies remains a problem of collecting, processing and information control. The majority of users in the Russian segment of the Internet use the existing retrieval services while the information collecting. According to the information on November, 02 2015 the most popular information search services are as follows: Google, Yandex, Mail, Rambler which account for 95% of the user queries. The distribution of the user queries number can be given in the follow way (Fig. 1).

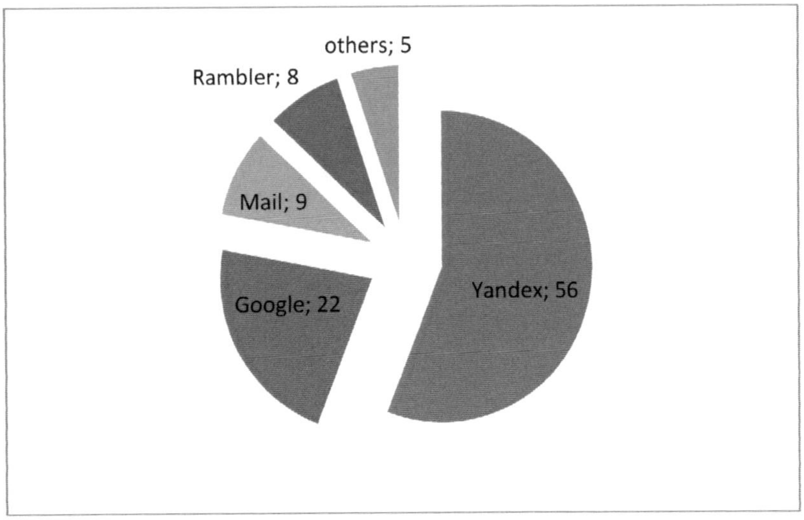

Figure 1. The share of retrieval queries in Russian segment of the Internet.

However, it should be noted that these services give good results while working with general subjects [1]. But if you need to retrieve specialized information you may have difficulties. Also the problem of multilingual representation of information in the Internet can be correctly solved in these systems. The retrieval services of the general purpose retrieve only in the language set the specified retrieval query was conducted. However, when searching the highly specialized personalized information one can arrange a multilingual retrieval procedure.

To solve the described above problems it is offered to use the existing technologies and approaches but with the emphasis on the processing of subject-oriented information.

First, it is offered to use the proven technology of the information control systems implementation based on the multi-agent approach.

The considered structure of interaction among agents of the offered multi-agent system can be shown as follows (Fig. 2).

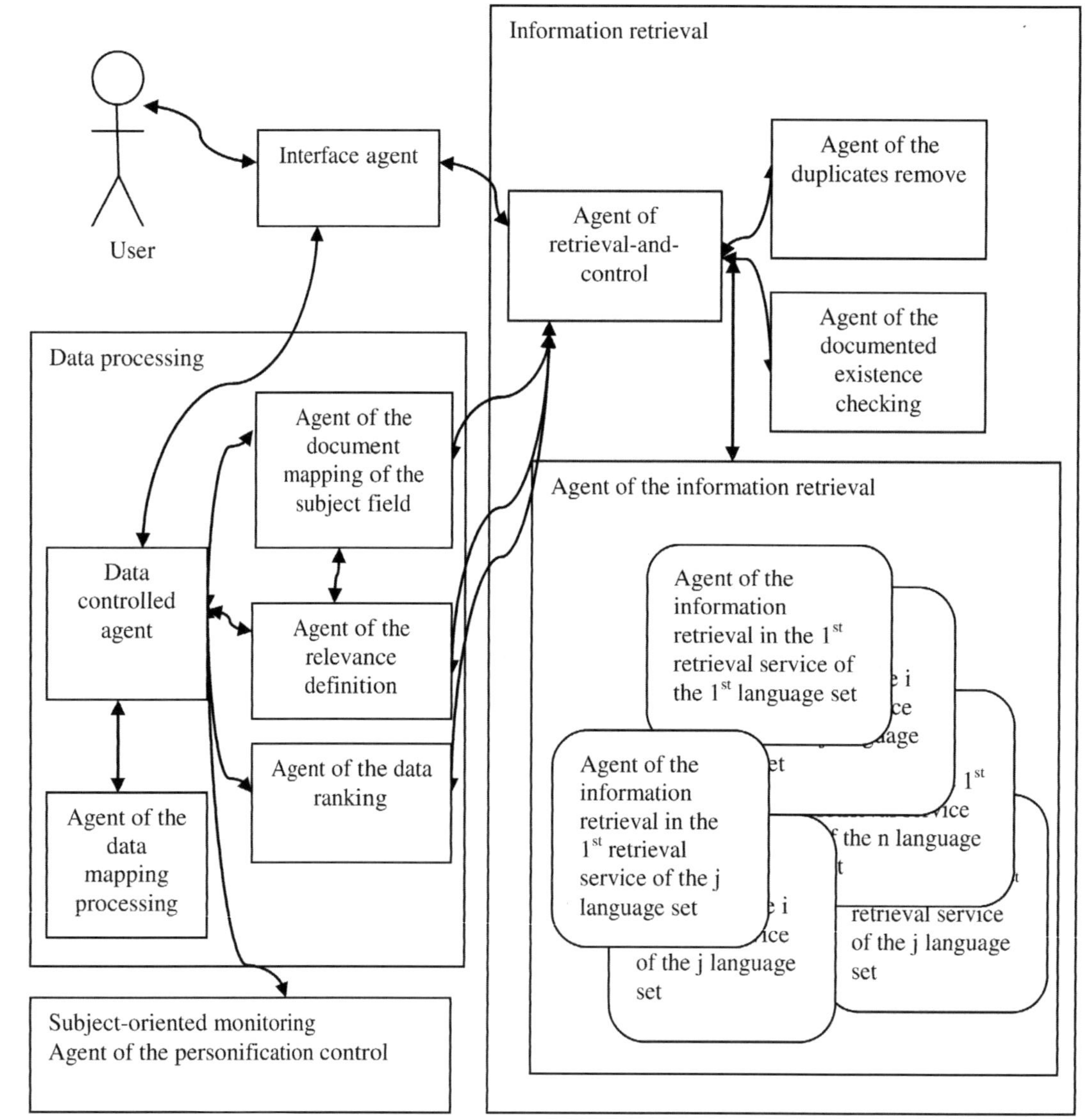

Figure 2. General scheme of the offered multiagent system

As one can see in the structure of the considered system it is necessary to design three logically-related software modules [2]:

1. The first module is responsible for the information quality in the Internet retrieval (formation of multilingual subject-oriented collections). In addition, it is necessary this module to define the relevance of the documents from a thematic catalogs (an agent of the relevance determination). Using the algorithms of the relevance determination it is possible to show that some of the documents are in some way "like a more relevant queries", and some of them are "less relevant". Thus we have a similar complex problem for the user associated with the problem of the documents delivery (an agent of the data ranking). A significant problem that this module solves is the problem of identifying arbitrary relevant documents (documents from the related subject fields). Therefore, a user must take into account the possibility of falling into the resulted sample of the subject areas. If you know the proximity of these fields, you should take into account a user preference of the system and solve the problem of inclusion or exclusion of the documents from the related subject fields into the resulted sample (an agent of the document mapping of the subject field who works closely with the agent of the relevance definition). In addition, only a part of the whole document can be in the selected text of the subject field, for example textbooks of a general purpose, collections of papers, etc. Therefore it is necessary to make a decision to ensure this user with only the necessary information.

2 The second module is responsible for the thematic collection control from the point of view of the user of this system. The main problems here is information processing from adjacent subject fields of the thematic collections (for example, the information on systems analysis and higher mathematics), as well as the storage and processing of the multilingual information in the framework of subject-oriented collections. Also the problem of the system response ranking to a user inquiry is also significant while this subsystem realization [3].

3. The third module is responsible for the subject-oriented monitoring of the user preference of thematic collections and the provision with the personalized navigation support and personalized data based on the identified thematic preferences. Providing users with the information collections of the personalized navigation menu of links into the pages that are close to their thematic preferences, the time required to retrieve the relevant information will; reduce, and the number of pages they view will increase.

The proposed solution should improve the usability of users with information resources and will serve as the additional motivation for them to visit these services more often: it will lead to the formation of a more constant audience of the thematic-oriented collections in the Internet.

References

[1] Zelenkov P. V., Karaseva M. V., Kovalev I. V. Models and methods for optimization of collecting and processing information // Vestnik of SibSAU. – Vol. 2. – 2012. – p. 69-72.

[2] Karaseva, M.V., Bachurina, E.P., Zelenkov, P.V., Brezitskaya, V.V. Information search module based on multilinguistic thesauruses (2015) IOP Conference Series: Materials Science and Engineering, 70 (1), art. no. 012011. http://www.scopus.com/inward/record.url?eid=2-s2.0-84921447476&partnerID=40&md5=728ef39ad283268034a48b7245b47dcf

[3] Kovalev I. V., Karaseva M. V., Leskov O. V. Components of information support for multilingual adaptive-learning technology / Scientific technical journal "Control systems and information technologies", Scientific book, 2009, № 1.3 (35). – S. 360 - 363.

IOP Conf. Series: Materials Science and Engineering **122** (2016) 012019 doi:10.1088/1757-899X/122/1/012019

Calculation and optimization of parameters in low-flow pumps

E M Kraeva [1] and I S Masich [1,2]
[1] Siberian State Aerospace University named after academician M. F. Reshetnev, Krasnoyarsk, Russia
[2] Department of Informatics and Mathematical Modeling, Krasnoyarsk State Agrarian University, Krasnoyarsk, Russia

E-mail: i-masich@yandex.ru

Abstract. The materials on balance tests of high-speed centrifugal pumps with low flow rate are presented. On the bases of analysis and research synthesis, we demonstrate the rational use of impellers of semi-open and open types providing high values for energy parameters of feed system of low-flow pumps.

Currently, work is continuing on the improvement of pumps for various purposes, especially centrifugal pumps with a semi-open impeller, where impeller type is selected mainly for constructional reasons and only sometimes for technology. Impellers of high-speed centrifugal pumps (HSCP) have small size and are well assembled with a high-speed small-sized axis gas turbine. This predetermines their widespread use in energy installations, primarily in turbo pump assemblies of liquid fuel rocket engines for systems of orientation, stabilization and brake propulsion.

Because the theoretical methods for calculating the jet-vortex flow around lattices of half-open impellers are developed for the rational design impellers for exactly such centrifugal pumps, so when we select the best constructive solutions for an entire pump, paramount importance until now belongs to experiments that allows to specify the basic theoretical principles.

HSCP with small flow rate $\dot{V}$ at the angular velocity of the rotor ω up to 10 000 rad/s are widely used as a part of turbopump assembly of liquid fuel rocket engines with low thrust and in energy installations of a spacecraft. This is due to a wide range of changes in pump mode parameters. For example, when the angular velocity is from 3000 to 10000 rad/s, the value $\dot{V}/\omega$ is usually less than $(\dot{V}/\omega) = 1 \cdot 10^{-6}$ m^3 which is the maximum permissible for centrifugal pumps with a closed impeller [1].

However, in the design of low-flow centrifugal pumps (LFCP) for aerospace applications, techniques and principles for optimization of flowing part are commonly used that are recommended for full-size centrifugal pumps, usually with a closed-type impeller [2]. This significantly limits mode parameters of pump units.

Hydrodynamic analysis [3] of viscous fluid flow around a lattice of LFCP profiles shows a significant displacement of the flow core from the channel walls and narrowing of the flow cross section in interscapular channels up to 80%. This leads to sharp decline in energy parameters of LFCP down to disc pump parameters [1]. Such blade pumps have a low value of the specific speed ratio:

$$n_s = 193.3\frac{\omega\sqrt{\dot{V}}}{H^{0.75}}, \quad (1)$$

where ω is angular velocity of the rotor, rad/s;
$\dot{V}$ is volume flow rate of the working fluid, m^3/s;
H is pressure of the working fluid, J/kg.

The value n_s <50 in pumps with low flow rate levels (50 ... 300 cm^3 / s) is mainly determined by the angular velocity and pump pressure. Such blade pumps, compared to pumps with high values n_s, have an increased proportion of hydraulic, mechanical and consumable losses. The proportion of these losses in the total energy balance is great and can be up to 70% of power consumption [1].

When designing flow part of a centrifugal impeller with n_s <50, small width can be obtained at the output of impeller (sometimes less than 1 mm), which is technologically difficult to implement and often designer increases the channel width of the impeller considering the boundary layer thickness. Such an increase in the channel width leads to a high degree of diffusivity $\overline{W}$ and a significant reduction in the energy parameters of the pump [4]. With increasing of viscosity, addition to reducing efficiency and impeller pressure, there is displacement of the greatest value of the efficiency to the region of lower flow rate. Occurs as though locking effective section of the impeller channel due to the growth of the boundary layer thickness, which is consistent with the calculation model presented in [3; 4].

In [5] on the basis of generalization of data presented in the works on HSCP with low flow rate in various organizations of aerospace industry, authors noted the need for a detailed study of HSCP with semi-open impellers as a promising high-speed pump at a value n_s <50 and $\dot{V}/\omega$ < $5\cdot10^{-7}$ m^3. In improving the flow part of HSCP with low flow and optimizing constructive solutions, great importance is experiments with carrying out the balance tests and analysis of the results.

The famous technique of S.S. Rudnev for balance testing large sized centrifugal pumps have allowed with some interpretations in a number of studies to realize the carrying out of balance tests with HSCP with impellers of closed, semi-open and open type according to the patent [6].

Analysis of the energy balance of HSCP with semi-open and open impellers has allowed to reveal features of the balance of energy loss in a pump with allocation the components of certain types of losses. Along with the balance tests, the energy characteristics of a centrifugal pump were taken with change in a wide range of constructive and mode parameters that are of practical and scientific interest. For example, the end clearance between the blades and the pump body were varied in different combinations of impellers from 0.25 mm to 3 mm. For purity of experiments, testing HSCP with different types of impellers were carried out with the simplest sealing system without the use of impellers, recirculation systems and so on, the value of the flow rate was discrete changing from 0 to 1.5 $\dot{V}_n$ through each 30 ... 50 ml/s. A nominal value of flow rate was equal to a value corresponding to the largest total efficiency of the pump.

IOP Conf. Series: Materials Science and Engineering **122** (2016) 012019 doi:10.1088/1757-899X/122/1/012019

Comparative tests and assessment of the balance of energy losses in HSCP with impellers of different types showed bias of mode of the most hydraulic pump efficiency to the region of lower flow rate. It is thus noticed the redistribution of certain types of energy loss. Thus, along with decreasing mechanical efficiency η_m^k of an impeller, consumable efficiency η_p was increased. For a semi-open impeller, that is identical to closed impeller by flowing parts, all this generally led to change of the efficiency of the pump from 0.535 to 0.525. When ω = 1047 / s for the mode with the greatest efficiency of the pump with semi-open impeller, efficiency components were as: full η_n = 0.519; hydraulic η_g = 0.699; consumable η_p = 0.924; mechanical impeller η_m^k = 0.828; mechanical parts of the stator η_m^s = 0.969.

The absence of coating disc and front gap seals significantly reduced leakage rate. At the same time mechanical losses of the impeller increased due to friction of end bladed surface. The hydraulic efficiency slightly decreased down to 0.699 due to energy loss in friction of fluid flowing in the channels of the half-open impeller. It should be noted that the value of the complex ($K_z\eta_g$) for a HSCP with the half-open and open impeller at β_{2l} near to 90° remains constant with a change in the flow rate.

Hydraulic characteristics of a centrifugal pump with an open and semi-open impeller have the best value in the region of low flow rate. The hydraulic efficiency of an open centrifugal pump is 0.75 - 0.78 in the entire range of low feed

$$\dot{V}/\omega \le 1.5 \cdot 10^{-7}\ \mathrm{m}^3,$$

with a substantial increasing consumable efficiency up to 0.997 (there are practically no leakage). However, due to friction on the end surfaces of blades of the impeller, there are observed a significant loss, the value of which has increased up to the value η_m^k = 0.7 because of removal the part of the second disk. Herewith, the average level of energy parameters, which are similar for HSCP with closed impellers, at

$$\dot{V}/\omega \le 1 \cdot 10^{-7}\ \mathrm{m}^3$$

was not only reached, but also was surpassed.

Losses in the drainage devices was determined integrally over the entire channel of drain in HSCP as the amount of loss in the spiral collector L_s and the conical diffuser $L_{k.d.}$. When testing various types of impellers, the hydraulic efficiency of drainage devices in centrifugal pumps had only little dependence on the type of the impeller and in a greater extent was determined by the operating mode of the pump. And this efficiency was decreases with increasing flow rate [1]. In the analysis of the energy characteristics of drainage in HSCP, the methodology of B.I. Borovsky [7] is accepted and the coefficient of drain losses is calculated

$$\xi_o = \frac{2L_o}{C_{2u}^2}.$$

The value of this coefficient for a wide range of pump flow rate $\dot{V} = (0.2...1.2)\dot{V}_n$ equals to $\xi_{o.p.}$ = 0.2 ... 0.3, which is consistent with the data for high-flow pumps [7] in the region of low feed and exceeds their values at nominal and high flow rate. Herewith, ξ_o = 0.25 ... 0.3 for calculated mode, and that was accepted for further study of LFCP. It should be noted that for all tested impellers the minimum for coefficient of drain loss coincides with the maximum for hydraulic efficiency.

In pump theory, a complex issue is determination of the angle of flow deflection from the direction of the blades at the output of the impeller $\delta = \beta_{2l} - \beta_2$ and the associated rate of influence of a finite number of blades K_z [7].

$$K_z = \frac{H_T}{H_{T\infty}} \tag{2}$$

Directly on the characteristics of the pumps tested, the product

$$K_z\eta_e = \frac{H}{H_{T\infty}} \tag{3}$$

can be determined. And conversely, it is easily to determine pump pressure from the known value of product ($K_z\eta_g$).

The results of generalization of known data on the value ($K_z\eta_g$) for a centrifugal pump with various types of impellers allowed to accept the value $K_z\eta_g = 0.62...0.68$. Herewith the materials of work [7] are used, that summarizes the tests of high-flow pumps with a closed impeller ($\beta_{2l} < 90°, n_s = 60...180$; z = 8...12) under changing

$$\overline{D}_1 = {D_1}/{D_2} = 0.8...0.3.$$

And the tests of pumps with n_s <50, including HSCP with low flow rate in systems of thermal control in a spacecraft [2]. It should be noted the identity of the recommendations on the results of all work about the values $K_z\eta_g$ which is constant at $\overline{D}_1 = 0.3...0.55$.

A large number of semi-empirical methods and dependencies are known, that allow to estimate the value K_z for a given class of pumps. However, they consider only closed impellers and are not applicable for calculation of pumps with semi-open impellers. Balance testing of centrifugal pumps with semi-open and open impellers allows to analyze the impact of design and operational factors individually on the value K_z and the hydraulic efficiency of the pump.

With a decrease in $\overline{D}_1 < 0.2$, the complex $K_z\eta_g$ decreases only for HSCP with a closed impeller due to the increase in hydraulic losses at flow in the channels [7]. It should be noted that with the growth of the working fluid viscosity, this value comes at a larger value $\overline{D}_1$ [3]. That is determined by increasing the thickness of the boundary layer displacement at flow in narrow channels of a closed impeller.

From the foregoing it follows that the rational application of HSCP with a half-open impeller is acceptable at $\overline{D}_1 = 0.1 \ldots 0.55$. It should be considered dependence ($K_z\eta_g$) from the angle β_{2l}. For an impeller with β_{2l} increased up to 90°, a value ($K_z\eta_g$) is high.

The peculiarity of the working process in channels of semi-open and open impellers at vortex interaction with the blades of radial fluid flow causes a stable value $K_z\eta_g$ over a wide range of variation of angular speed of the rotor of the pump (Figure 1). For impellers of semi-open and open types in the whole region of change the value $\overline{D}_1 < 0.55$, there is a consistently high value $K_z\eta_g = 0.62 \ldots 0.68$. That determines preferential using these impellers in HSCP at $\overline{D}_1 < 0.25$. And it is typical for high-pressure pumps in power installations of a spacecraft.

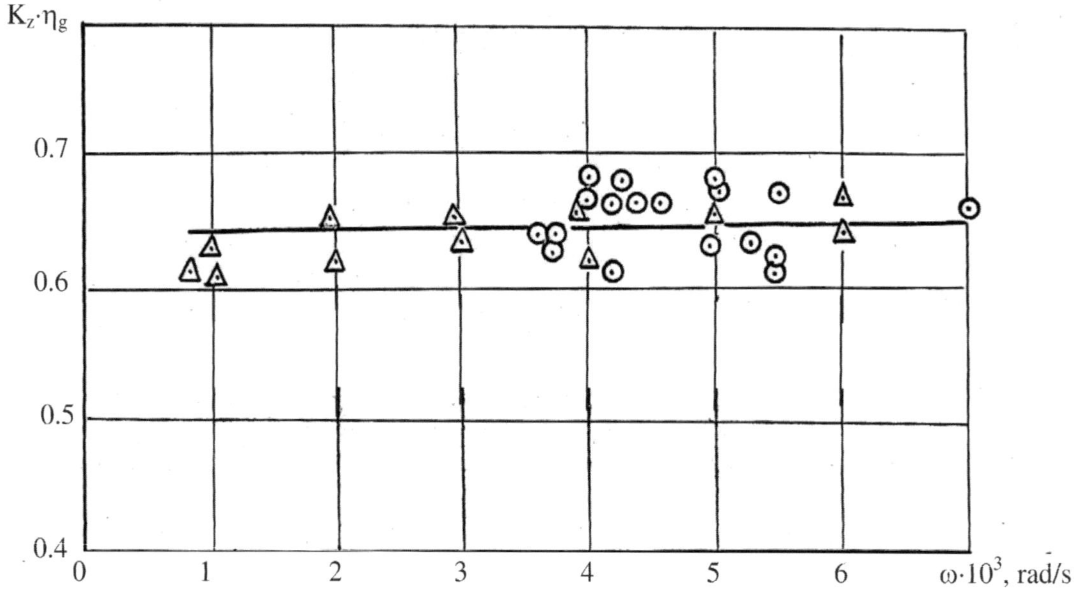

Figure 1. Influence of the angular velocity on the change of the complex $K_z\eta_g$ for impellers of semi-open and open types :

 - tests obtained in SibSAU

 - tests obtained in other researches

IOP Conf. Series: Materials Science and Engineering **122** (2016) 012019 doi:10.1088/1757-899X/122/1/012019

Experimental data for open and semi-open types of an impeller are approximated at $\beta_{2l} = 90°$ by dependence of the form:

$$K_z = 0.2 + 0.7(1 - \overline{D}_1^{12}) \quad (4)$$

or, taking into account the value $\beta_{2л}$ the dependence has been obtained:

$$K_z = 0.2 + 0.67\beta_{2l}^{0.15}\left[1 - \left(\frac{\overline{D}_1}{\sin\beta_{2l}}\right)^{12}\right] \quad (5)$$

where β_{2l} is tilt angle of the blade at the outlet, in radians.

The impact of the relative diameter $\overline{D}_1$ on the value K_z is consistent with similar data for closed impellers [7] and occurs at low density lattices. At the usual number of blades z = 4 ... 8 the impact of the relative diameter of an impeller occurs in area $\overline{D}_1 > 0.6...0.7$. Increasing the value K_z can be achieved by increasing the number of blades z. However, it clutters through passage section of the impeller, and this leading to increased hydraulic loss. The optimum number of blades providing the largest value K_z can be found from the dependence [5; 8]

$$z = \frac{\pi\tau\sin\left(\frac{\beta_{1l} + \beta_{2l}}{2}\right)\left(1 + \overline{D}_1\right)}{\left(1 - \overline{D}_1\right)}, \quad (6)$$

where τ is density of the lattice of profiles.

The analysis of the balance tests for centrifugal pumps with semi-open impellers confirmed the admissibility of dependence (5) in calculation of the value K_z for the optimized (on density) circular lattice of profiles at the flow part in the channels of the impeller and also in calculation of its design features. It should be noted that the results obtained from the dependence of the coefficient K_z on the angle β_{2l}, the values $\overline{D}_1$, τ and z are acceptable for operation mode of HSCP at

$$C_{2m}/U_2 \leq 0.1.$$

That corresponds to the parameter

$$\dot{V}/\omega \leq 5 \cdot 10^{-7}\ \text{m}^3.$$

The analysis of the balance tests for HSCP with half-open and open impellers has shown the value $\eta_g = 0.8...0.85$. That meets the modern requirements for energy excellence of pumps (Figure 2). For comparison, there the experimental data are presented for HSCP with closed impellers. These data are described by the dependence taken from the work [1]

$$\eta_g = 0.38 \cdot n_s^{0.21}. \tag{7}$$

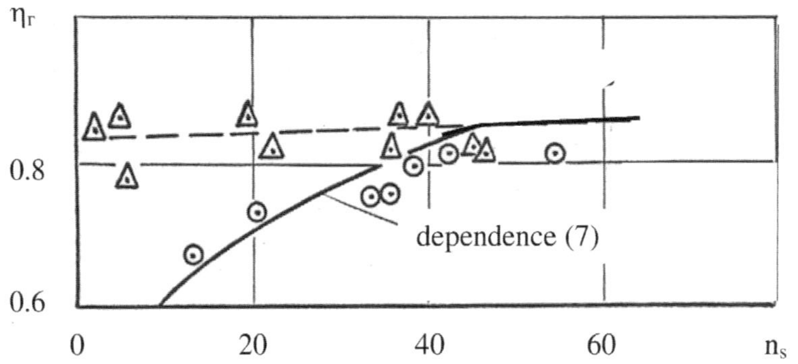

Figure 2. Dependence of the hydraulic efficiency for HSCP with various types of impellers on speed coefficient:
▲ – semi-open and open impeller; ⊙ - closed impeller.

Thus, at the low value n_s, and therefore, at the value $\overline{D}_1 < 0.25$, in order to obtain a high pump efficiency, it is recommended to apply an impeller of semi-open or open types because its power characteristics exceed those of pumps with closed impellers.

Special attention in the study of HSCP with the semi-open and open impellers was given to assessing the impact of the value a_1 of a lateral gap between the impeller and the pump body on change the energy characteristics of the pump to optimize sizes of gaps, providing not only the highest settings, but also processability of HSCP.

The tests with analysis of the impact were conducted on a number of pumps with impellers with different geometry at various angular rotor speeds from 500 to 4400 rad/s. The value a_1 of the end gap changed from 0.15 to 3 mm on both sides simultaneously, and also at a fixed value of one. Furthermore, we tested impellers with variable width of the blade at the outlet of the impeller for the series of constant gap values ($b_2 = 2.7...9.5$ mm). That has allowed to experimentally verify the condition to minimize blade width at the output of the impeller according to the theoretical relation in the work [8] from the condition of conservation of the flow core in the channels of the impeller by the oncoming jets from the end edges of the blades.

Dispersion of the experimental points for dependence (8) is $\sigma = 0.023$ and for dependence (9) is $\sigma = 0.00455$, which is quite satisfactory. In the result of testing the limit values a_1/b_2 are verified. HSCP with flowing part parameters, which do not corresponding to the received ratios, are characterized by low energy parameters.

We consider the rational use of centrifugal pumps. It is shown that the centrifugal high-speed pumps with closed impellers at $\overline{D}_1 < 0.25$ have low efficiency due to the high proportion of hydraulic losses. That allows us to recommend their use in regimes

$$\dot{V}/\omega \geq 1.5 \cdot 10^{-7}\,\mathrm{m}^3.$$

Experimentally we studied how the main design parameters of pumps with semi-open and open impellers effect on the efficiency and the pressure coefficient. Taking into account the theoretical studies, recommendations have been obtained on the selection of optimal parameters providing improvement the overall level of energy performance of a centrifugal pump in the area of unit feed

$$\dot{V}/\omega \leq 5 \cdot 10^{-7}\,\mathrm{m}^3$$

down to zero.

Recommendations were developed for the calculation of energy parameters of HSCP with impellers of semi-open and open types. Experimental data confirmed the basic theoretical principles. The results of studies are presented in form of generalized dependencies that allow to widely use them in engineering practice.

Acknowledgments

The results were obtained within the project No. 346 of the state task No. 2014/211 of the Ministry of Education and Science of the Russian Federation.

References

[1] Kraev M V, Lukin V A and Ovsyannikov B V 1985 *Low-flow pumps of aviation and space systems* (Moscow: Mashinostroenie) p 128

[2] Bobkov A V 2003 *Centrifugal pumps in thermoregulation systems of space vehicles* (Vladivostok: Dal'nauka) p 217

[3] Kraeva E M 2010 Theoretical analysis of flow around lattice of profiles by viscous fluid flow in low-flow centrifugal pump *Vestnik SibSAU* **1(27)** 96-99

[4] Kraeva E M 2010 Calculation of energy parameters of high-speed centrifugal pumps of low rapidity *Aeronautical engineering* **1** 48-50

[5] Kraev M V, Kishkin A A and Sizykh D N 1998 *Hydrodynamics of low-flow pumps* (Krasnoyarsk: Siberian Aerospace Academy) p 157

[6] Kraev M V, Kishkin A A and Melkozerov M G 2003 The method on balance tests of impeller: *patent for invention* (Russia) N 2217724 Bulletin **33**.

[7] Borovsky B I, Ershov N S, Ovsyannikov B V 1975 *High-speed blade pumps* (Moscow: Mashinostroenie) p 335

[8] Kraeva E M 2004 Jet-vortex model for hydrodynamic of flow part of centrifugal impeller *Vestnik SibSAU* **5** 173-179

To the question about the states of workability for automatic control systems with complicated structure

P A Kuznetsov[1], I V Kovalev[2], V V Losev[1], A O Kalinin[1], A V Murygin[2]

[1]Siberian State Technological University
82, Mira Av., Krasnoyarsk, 660049, Russian Federation
[2]Siberian State Aerospace University named after academician M. F. Reshetnev
31, Krasnoyarsky Rabochy Av., Krasnoyarsk, 660037, Russian Federation

E-mail: zelenkow@rambler.ru

Abstract. The article discusses the reliability of automated control systems. Analyzes the approach to the classification systems for health States. This approach can be as traditional binary approach, operating with the concept of "serviceability", and other variants of estimation of the system state. This article provides one such option, providing selective evaluation of components for the reliability of the entire system. Introduced description of various automatic control systems and their elements from the point of view of health and risk, mathematical method of determining the transition object from state to state, they differ from each other in the implementation of the objective function. Explores the interplay of elements in different States, the aggregate state of the elements connected in series or in parallel. Are the tables of various logic States and the principles of their calculation in series and parallel connection. Through simulation the proposed approach is illustrated by finding the probability of getting into the system state data in parallel and serially connected elements, with their different probabilities of moving from state to state. In general, the materials of article will be useful for analyzing of the reliability the automated control systems and engineering of the highly-reliable systems. Thus, this mechanism to determine the State of the system provides more detailed information about it and allows a selective approach to the reliability of the system as a whole. Detailed results when assessing the reliability of the automated control systems allows the engineer to make an informed decision when designing means of improving reliability.

Keywords: reliability, states, faultlessness, workability.

I Introduction. Reliability of the systems with complicated structure is the sphere of research with wide range of various methods and computations. According modern approach reliability is the property of the object, allowing it to keep the entire parameters characterizing the ability to perform the required functions constant in time. [1, 2, 3, 4].

Reliability of each element can be expressed in value of P_i – possibility of faultless operations, and state – by the traditional logical variable S_i, taking the value 1 if the element serviceable and 0 in case of failure.

Random event, expressed by the number of R_i. taking the value in interval (0…1), determines the switching of states. Thus, for determining of element state is possible to use expression:

$$S_i = 1, \text{ if } R_i < P_i \quad (1)$$
$$S_i = 0, \text{ if } R_i > P_i \quad (2)$$

The state of entire system expresses by the calculating of the logical formula of the system, which arguments is the states of its elements. Traditionally these states called "workable" and "unworkable". This dualistic approach allows to apply usual Boolean logic to the calculation of the current state of the system.

II Safety point of view to the states of the system The very set of states is not something absolutely true and certain. It is possible to expand the idea of the state of the elements / systems. Indeed, apart from the usual idea of the state as a "workable" and "unworkable", there are other classifications of states [6].

In accordance with the approach of the safety of the systems, increasing the reliability of the elements by introducing structural and temporal redundancy, applying interchangeability, maintainability and other measures to increase the reliability of a complex system, fault tolerance is provided. But it is characteristic of complex systems is the possibility of multiple combinations of failures, each of which is improbable, but the amount of such incredible states accumulates enough to hit the system in a dangerous state - a state in which there is damage to "large scale."

In [7] presented the basic definitions that are used in this issue. But along with that assess the state of security of the system, a plurality of states can be considered in assessing the health of the system.

III System of states calculating For the operating of the various states specific state calculating system, based on Boolean logic, is required. The difference of the proposed system is a greater number of possible values of state variables.

To illustrate the versatility of this approach consider several states V_1. V_2, V_3, V_4, which can exist as a system and its individual elements.

State V_1 - inoperable. In this state, the element is not fully perform its function, and all parameters are not as desired.

State V_2 - defective. In it the element performs its function, but with some kind of error, mistake.

State V_3 - workable. In it meet the requirements of normative-technical documentation and design values of all technical parameters of elements, qualities which characterize their ability to perform a given function. Values of other parameters may not match the documentation.

State V_4 - good state. In it the element meets all the regulatory, technical and engineering documentation [1, 8].

Elements of the system can be interconnected in various ways. The complexity of the system can provide increase its reliability by applying the principle of redundancy. By redundancy is understood the application of certain technical means to ensure the health facility in case of failure. [9] In systems with redundant primary and backup isolated elements: the first is an element of the structure of the object, whose failure in the absence of a reservation leads to a loss of efficiency of the object, the second - a component designed to provide health facility in case of failure of the basic element [10, 11].

IV Tables of the state of serial and parallel connected elements Based on the foregoing, we describe how to calculate the state of the different connected elements. When elements connected serial the logic function is similar to the conjunction of elements and the state table is as follows:

Table 1. State of serial connected elements.

S_1	S_2	S_1 & S_2
V_1	V_1	V_1
V_1	V_2	V_1
V_1	V_3	V_1
V_1	V_4	V_1
V_2	V_1	V_1
V_2	V_2	V_2
V_2	V_3	V_2
V_2	V_4	V_2
V_3	V_1	V_1
V_3	V_2	V_2
V_3	V_3	V_3
V_3	V_4	V_3
V_4	V_1	V_1
V_4	V_2	V_2
V_4	V_3	V_3
V_4	V_4	V_4

In a case of parallel connection states calculating according the next table

Table 2. State of parallel connected elements.

S_1	S_2	$S_1 \| S_2$
V_1	V_1	V_1
V_1	V_2	V_2
V_1	V_3	V_3
V_1	V_4	V_4
V_2	V_1	V_2
V_2	V_2	V_2
V_2	V_3	V_3
V_2	V_4	V_4
V_3	V_1	V_3
V_3	V_2	V_3
V_3	V_3	V_3
V_3	V_4	V_4
V_4	V_1	V_4
V_4	V_2	V_4
V_4	V_3	V_4
V_4	V_4	V_4

Each element of the system can move from state to state with a certain probability. Therefore, the calculation of system reliability can be performed on existing states, and on the probability of hitting the elements in these states.

V Example of the state calculation for the system of differently connected elements

System, depicted on Fig.1 illustrates an example of the calculation. This system consists of some elements M1, M2, M3 are connected in series and in parallel

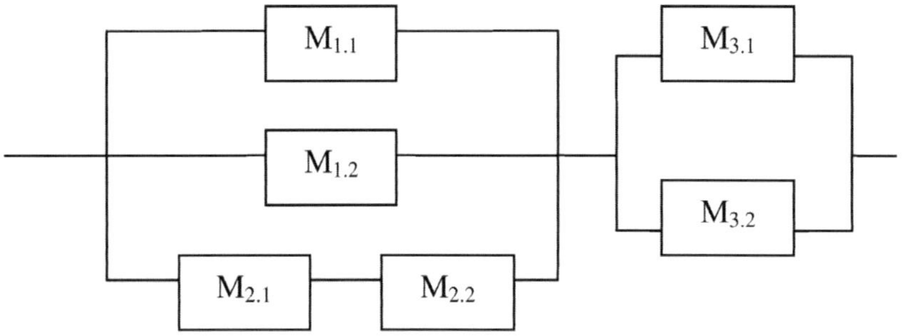

Figure 1. Structure of the example system

For the calculating of the state of the system probability of each state of each element is required.

Table 3. Probabilities of each state of each element

	P_{s1}	P_{s2}	P_{s3}	P_{s4}
M_1	0,001	0,05	0,45	0,4990
M_2	0,0001	0,02	0,5	0,4799
M_3	0,0005	0,03	0,6	0,3695

P_{sn} – probability of n-th state.

Analytically, the probability of the system in any state P_{tot} expressed by the formula

$$P_{tot} = (1\text{-}(1\text{-}P_{m1})^2\cdot(1\text{-} P_{m2}^{\ 2}))\cdot(1\text{-}(1\text{-} P_{m3})), \quad (3)$$

Results of calculation presented in table 4

Table 4. Probabilities of each state of the system

P_{s1}	P_{s2}	P_{s3}	P_{s4}
0,000002	0,005784	0,483825	0,5104

To illustrate the interaction of the system states run the simulation to generate a set of random states.

We generate the set of states and calculate the combination of states in accordance with our logic.

Table 5. Results of imitation modeling of the states

State of $M_{1.1}$	State of $M_{1.2}$	State of $M_{2.1}$	State of $M_{2.2}$	State of $M_{3.1}$	State of $M_{3.2}$
2	3	4	3	2	3

Logical function of the system

$$(M_{1.1} \vee M_{1.2} \vee (M_{2.1} \wedge M_{2.2})) \wedge (M_{3.1} \vee M_{3.2}),$$

$$M_{2.1} \wedge M_{2.2} = V_4 \wedge V_3 = V_3, \quad (5)$$

Parallel connected elements M_2 will have state V_3, workable.

$$M_{1.1} \vee M_{1.2} \vee (M_{2.1} \wedge M_{2.2}) = V_2 \vee V_3 \vee V_3 = V_3, \quad (6)$$

First parallel connection will result workable state.

$$M_{3.1} \vee M_{3.2} = V_2 \vee V_3 = V_3, \tag{7}$$

Second will have workable state too.

Thus

$$(M_{1.1} \vee M_{1.2} \vee (M_{2.1} \wedge M_{2.2})) \wedge (M_{3.1} \vee M_{3.2}) = V_3 \wedge V_3 = V_3,$$

System will be in the workable state.

VI Conclusion This differentiated approach to safety has many advantages. First, the traditional approach, defining only two states of the system - workable and unworkable, too rough. Knowing that the system is operational, we gain a sense of confidence in it, which is often false, since system with workable condition could have important non-functional parameters been out of normative.

Secondly, there is the possibility of a more detailed examination of states. Selectivity in the approach to system performance allows it accurately select desired items, and subsequently, the estimated state of the system. After all, the system is often classified according to their importance in a wide range. And a certain condition may not be satisfactory for a single system, for example, control the spacecraft, but satisfactory for another, such as household system.

Thus, definition, provided by this paper determines the more detailed state of the system. gives more information about it and allows a selective approach to the reliability of the system as a whole.

References

1. Kurochkin Yu.A., Smirnov A.S., Stepanov V.A. *Nadezhnost' i diagnostirovanie tsifrovykh ustroystv i sistem* [Reliability and diagnostics of digital systems]. St.Petersburg. St.Petersburg University Publ. 1993. 320 p.

2. Aleksandrovskaya L.N., Aronov I.Z., Kruglov V.I., Kuznetsov A.G., Patrakov N.N., Sholom A.M.. *Bezopasnost' i nadezhnost' tekhnicheskikh sistem. Uchebnoe posobie*. [Safety and reliability of technical systems. Study guide]. Moscow. Logos Publ. 2004. – 280 p.

3. Sugak E.V, Vasilenko N.V. *Nadezhnost' tekhnicheskikh sistem. Uchebnoe posobie dlya vysshikh uchebnykh zavedeniy* [Reliability of technical systems. Study guide]. Krasnoyarsk. NII SUVPT Publ., 2000. – 608 p.

4. Losev V.V. [Reliability analysis of ACS] // Vestnik universitetskogo kompleksa: Sb. nauchn. Tr. Vol 11(25) VSF RGUITP, NII SUVPT Publ. 2008 p.16-19.

5. Kovalev I.V., Kuznetsov P.A., Zelenkov P.V., Shaydurov V.V., Bakhmareva K.K.. [To the question of the reliability of automated control systems with blocking protection modules]. *Pribory*. Vol 6. 2013 p.20-24 (In Russ.).

6. Ryabinin I.A.. *Nadezhnost' i bezopasnost' strukturno-slozhnykh sistem* [Reliability and safety of structural complex systems]. St.Petersburg. St.Petersburg University Publ. 2007. 276 p.

7. Fedorov Yu.N, *Spravochnik inzhenera po ASU TP: Proektirovanie i razrabotka. Uchebno-prakticheskoe posobie*. [PCS engineer reference: Design and development. Training and practical guide] Moscow, Infra-inzheneriya publ., 2015. – 929 p.

8. GOST 27.002-89 Nadezhnost' v tekhnike. Osnovnye ponyatiya. Terminy i opredeleniya [Reliability in the technology. Basic concepts. Terms and Definitions] (In Russ.) Available at: http://vsegost.com/Catalog/11/11290.shtml (accessed 12.02.2015).

9. Kuznetsov P. A. [On the analysis of the effectiveness of a fully redundant systems]// Vestnik SibGAU 2015 Vol 16 #2. P. 326-331.

10. Kuznetsov P. A., Beschastnaya N. A., Bakhmareva K. K., Antamoshkin O. A., Antamoshkin A. N. [Modification of the Volkovich`s-Michalevi`c method to optimize costs in the synthesis of fault-tolerant information systems]. *Vestnik SibGAU*. 2012. Vol 6(46). P. 97-100 (In Russ.).

11. Kuznetsov P. A. [Dependant failures in multifunctional ACS]// *Vestnik SibGAU*.2015 Vol 16 #1. P. 86-91.

12. Petrokas L.V. *Teoriya mashin-avtomatov i pnevmogidroprivodov* [Theory of automated machines and pneumatic drives] Moscow. Mashinostroenie, 1970. 334 p.

13. *Slovar' po kibernetike /Pod redaktsiey V. S. Mikhalevicha* [Cybernetics dictionary. Editor: V.S.Mikhalevich] Kiev: Glav. red. Ukr. Sov. Jenciklopedii im. M.P. Bazhana, 1989. 751 p.

14. A. Akopov. Imitacionnoe modelirovanie. Uchebnik i praktikum – M.: Jurajt, 2015. – 390 p.

15. N.N. Pan'gina, A.A Pan'gin. Statistical modeling. Monte-Carlo method. // *Komp'yuternye instrumenty v obrazovanii. 2002. №5*. p. 30-43.

IOP Conf. Series: Materials Science and Engineering **122** (2016) 012021 doi:10.1088/1757-899X/122/1/012021

Method to eliminate the impact of magnetic fields on the position of the electron beam during EBW

V D Laptenok[1], A A Druzhinina[2], A V Murygin[3], Yu N Seregin[4]

[1]Professor, Siberian State Aerospace University named after Academician M.F. Reshetnev, 31 "Krasnoyarskiy Rabochiy" prospect, Krasnoyarsk, 660037, Russia

[2]Associate professor, Siberian Federal University, 79 Svobodniy prospect, Krasnoyarsk, 660041, Russia

[3]Professor, head of the chair of information control systems, Siberian State Aerospace University named after Academician M.F. Reshetnev, 31 "Krasnoyarskiy Rabochiy" prospect, Krasnoyarsk, 660037, Russia

[4]Associate professor, Siberian State Aerospace University named after Academician M.F. Reshetnev, 31 "Krasnoyarskiy Rabochiy" prospect, Krasnoyarsk, 660037, Russia

E-mail: avm514@mail.ru

Abstract. The paper presents the approximate formulas for calculating the deflection angle and the misalignment of the electron beam from the optical axis of the electron gun caused by the action of magnetic fields during the electron beam welding. Mathematical model of the effect of magnetic field induced by thermoelectric currents on the electron beam position in the process of electron beam welding of dissimilar materials is presented. The method of monitoring of the misalignment of the scanning electron beam and its mathematical model are proposed. Monitoring of the misalignment of the scanning electron beam is based on the processing of the signal of the collimated X-ray sensor directed to the optical axis of the electron gun by synchronous detection method. The method of compensation of the effect of magnetic fields by passing through the welded seam the currents which compensate thermoelectric currents is considered.

Introduction

One of the main causes of deflection of an electron beam during electron beam welding (EBW) is the effect of magnetic fields. These fields can be caused by the magnetization of the equipment, the residual magnetization of welded products, the action of various electromagnetic devices, thermoelectric currents produced by temperature gradients in some dissimilar materials [1]. Different ways of reducing the effect of magnetic fields are used. There are demagnetization of products, shielding of electron beam, and the compensation of magnetic field in the welding zone. Demagnetization of products can significantly reduce the noise level, but it may be re-magnetized. In addition, demagnetization of large parts is a labor-consuming and expensive process. Shielding of the electron beam by magnetic shield is the protection from external fields in the space of the shield location. This shield degrades possibility to monitor the process of EBW. Therefore, the compensation of magnetic field in the zone of magnetic field action is most appropriate [1].

Assessing the impact of the magnetic field

To assess the effect of magnetic field on the trajectory of the electron beam the formulas for calculating the deflection of the electron beam from the axis of the electron gun should be derived. Also, determination of the angle of the beam trajectory is an interesting, because it characterizes deflection of the weld seam from the plane of the joint.

The action of magnetic fields in the space between the electron gun and the work piece during EBW can be described with sufficient accuracy by the equations [2]:

(1) $$\psi = \frac{e}{mV}\int_0^z B(z)dz\,,\quad X = \frac{e}{mV}\int_0^z\int_0^z B(z)dzdz\,,$$

where ψ – angle of inclination of the trajectory of the electron beam relative to the optical axis of the electron gun z; X – displacement of the electron beam from the joint; e – electron charge; m – mass of the electron; V – electron speed, which depends on the accelerating voltage; $B(z)$ – projection of induction of the magnetic field in the plane of the joint as a function of the coordinate z.

The origin of coordinates (z=0) is the center of the focusing system.

The error of calculations according to formulas (1) for most cases occurring in practice doesn't exceed 2.5 %.

A mathematical model of magnetic fields induced by thermoelectric currents

When welding dissimilar materials the deflection of the electron beam from the joint can be caused by magnetic field of thermoelectric currents. According to Ohm's law, the current I in a closed circuit is directly proportional to the electromotive force and inversely proportional to the total resistance R of the entire circuit. Value of arising thermoelectric power E depends on the material of welded products and temperatures of hot ($T1$) and cold ($T2$) contacts. In a small range of the temperature the thermoelectric power can be considered as proportional to the difference of the temperature.

When welding two dissimilar materials 1 and 2 the total resistance of the entire circuit is the sum of the resistances R_1 and R_2 of these materials. Resistance of the conductor constant cross-section depends on the properties of the material of the conductor, its length and cross-section.

Thus, the thermoelectric current is determined by the formula

(2) $$I = \frac{E}{R} = \frac{\alpha_{12}S(T2-T1)}{\frac{\rho_1+\rho_2}{2}l_R}\,,$$

where α_{12} – thermoelectricpower coefficient; ρ_1, ρ_2 – resistivities of welded materials; l_R – length of the conductor; S – cross-sectional area of the conductor; $T1$ – temperature at the center of the weld where the circuit of thermoelectric currents is closed; $T2$ – melting temperature.

For determining the temperature $T1$ the equation of the temperature distribution in the infinite plate heated by a fast-moving line source without the heat transfer from the surface can be used [3]:

(3) $$T(x,y) = k_{12}T_{\max}\sqrt{\frac{v}{-y}}\exp\left(\frac{x^2v}{4ay}\right),$$

where k_{12} – coefficient characterizing the thermophysical properties of welded materials;

T_{max} – maximum temperature (evaporation temperature of welded materials); v – welding speed; a – coefficient of the temperature conductivity; x – coordinate across the joint; y – coordinate along the joint.

Equation (3) describes the temperature of an area which located behind the welding source.

Temperature at the center of the weld is determined by the formula

$$T = k_{12} T_{max} \sqrt{\frac{v}{-y}}. \quad (4)$$

Assume that the conductor with the thermoelectric current is an ellipse, as shown in Fig. 1. The length of the ellipse can be calculated from the approximate formula

$$l_R = \pi (a_y + a_x), \quad (5)$$

where $a_y = \frac{-y - y_w}{2}$ – semi-major axis of the ellipse; y_w – distance from the optical axis of the electron gun to the conductor with current (the length of the welding cavity along the axis y) ; a_x – semi-minor axis of the ellipse.

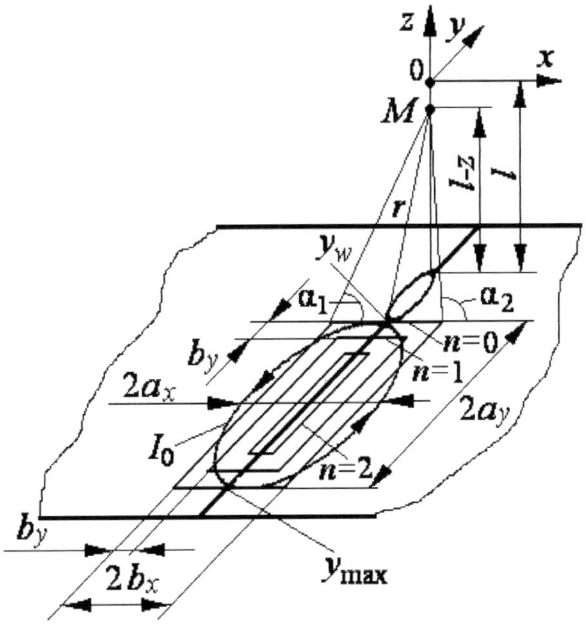

Figure 1. Schematic representation of the model.

Substitution of formulas 4 and 5 into expression 2, differentiation of the resulting expression with respect to y, equating of the derivative to zero allow to define coordinate y_{max} of closing the circuit of the maximal thermoelectric current I_0. Semi-minor axis of the ellipse should be set.

Values of circulating thermoelectric currents are defined by the expression

$$I_n = \frac{\alpha_{12} \delta b_x \left(k_{12} T_{max} \sqrt{\frac{v}{y_w + n b_y}} - k_{12} T_{max} \sqrt{\frac{v}{-y_{max} - n b_y}} \right)}{\frac{\rho_1 + \rho_2}{2} \cdot \pi \cdot \left(\frac{-y_{max} - y_w}{2} + a_x - 2 n b_y \right)}, \quad (6)$$

where b_y – width of the conductor with current; δ – thickness of the product; n= 0, 1, 2, 3….

Contours with thermoelectric currents are approximated by rectangles of length $(-y_{max} - y_w - 2nb_y)$ and width $(2b_x - 2nb_y)$, where n= 0, 1, 2, 3…. Components of magnetic field directed along the welding line, have the greatest impact on deflection of the electron beam from the joint. These magnetic fields are induced by line segments of current which are perpendicular to the weld.

According to the law of Biot-Savart-Laplace in the case when $\alpha_2=180-\alpha_1$, $\cos\alpha_1-\cos\alpha_2=2\cos\alpha_1$ the induction of magnetic field of the straight line segment of current having a finite length is described by the equation [4]

$$B = \frac{\mu_0 I}{2\pi r}\cos\alpha_1 = \frac{\mu_0 I}{2\pi r}\cdot\frac{b_x}{\sqrt{b_x^2 + r^2}}, \tag{7}$$

where b_x – half the length of the straight line segment of current; r – distance from the straight line segment of current to the point under consideration.

During the welding process conductors with a current are removed from the axis of the electron gun to distances (y_w+nb_y) and $(-y_{max} - nb_y)$, distance from the point under consideration M to the surface of the welded parts is $(l - z)$, where l – distance from the surface of the welded parts to the electron gun, as shown in Fig. 1. Distance from the straight line segment of current to the point under consideration is determined by formula

$$r = \sqrt{(y_w + nb_y)^2 + (l - z)^2}\,. \tag{8}$$

Thus, the equation (7) can be written as

$$B_n^+(z) = \frac{\mu_0 I_n}{2\pi\sqrt{(y_w + nb_y)^2 + (l - z)^2}} \times \frac{b_x - nb_y}{\sqrt{(b_x - nb_y)^2 + (y_w + nb_y)^2 + (l - z)^2}}, \tag{9}$$

$$B_n^-(z) = \frac{\mu_0 I_n}{2\pi\sqrt{(-y_{max} - nb_y)^2 + (l - z)^2}} \times \frac{b_x - nb_y}{\sqrt{(b_x - nb_y)^2 + (-y_{max} - nb_y)^2 + (l - z)^2}}. \tag{10}$$

Indices «+» and «–» show that vectors of magnetic induction are directed oppositely to each other.

Using expressions (6), (9) and (10), total deflection can be calculated according to equation (1).

For comparative assessment of calculated and experimental data the materials and their characteristics considered in paper [5] are used. Properties of welded materials are presented in Table 1. The results of calculations produced for welding stainless steel SS 304 and low-carbon steel showed that at an accelerating voltage of 60 kV, a welding speed of 40 m/h, the product thickness of 2.5 cm deflection of the weld from the joint at the depth of 6 mm is 0.6 mm. The calculation results agree well with experimental data obtained in the work [5].

Table 1. Properties of alloys

Alloy	$T2$ (K)	a (m^2/s) × 10^5	ρ (Ohm·m) × 10^{-7}
Stainless steel SS 304	1700	0.45	6.667
Low-carbon steel	1783	1.28	1.136

System of control and compensation of deflection caused by the action of magnetic field

In the interaction of the electron beam with the processed material electrons lose their energy as a result of braking. This process is accompanied by the excitation of the X-ray radiation, localized in a place where the electron beam interacts with the material.

The ability to control the deflection of the electron beam from the optical axis of the gun by the X-ray radiation from the zone of braking of electrons is based on the fact that the intensity of X-ray radiation depends from the surface area through which the radiation flux passes. Deflection of the beam from the axis of the electron gun caused by the action of magnetic fields, leads to a decrease of share of X-ray radiation passing through the crystal area of the sensor. Consequently, the intensity of X-ray radiation I_d measured by the sensor is reduced.

X-ray sensor includes X-ray detector and collimator, which is configured in the form of a blind with the slit oriented to the optical axis of the gun, as shown in Fig. 4.

For monitoring the beam deflection the technological scanning, which improves the quality of the welded joint is used.

A mathematical model of the X-ray sensor as an element having the frequency spectrum of the output signal at presence of the periodic signal of scanning of the electron beam is expressed by the relation [6]

$$I_d(t) = \frac{a_0}{2} + \sum_{n=1}^{\infty} \left(a_n \cos nt + b_n \sin nt\right), \quad (11)$$

where coefficients of the series are given by the expressions:

$$a_0 = \frac{1}{\pi} \int_{-\pi}^{\pi} I_d(t) dt, \quad (12)$$

$$a_n = \frac{1}{\pi} \int_{-\pi}^{\pi} I_d(t) \cos ntdt, \quad (13)$$

$$b_n = \frac{1}{\pi} \int_{-\pi}^{\pi} I_d(t) \sin ntdt. \quad (14)$$

Here, the intensity of X-ray radiation from the surface of the work piece is determined by expression

$$I_d = kk_1 U_0^2 Z I_e \frac{1}{\sigma_x \sqrt{2\pi}} \times \int_{-\infty}^{\infty} f_d(x) \exp\left(-\frac{\left(x - \varepsilon_0 - \varepsilon_m \sin \omega t\right)^2}{2\sigma_x^2} \right) dx, \quad (15)$$

where $f_d(x)$ – function of the view area of the collimator of X-ray sensor; k =1.5·10^{-9} V^{-1} – proportionality coefficient; k_1 – coefficient considering the orientation of the sensor in the space; I_e – electron beam current; Z – atomic number of the material of welded product; U_0 – accelerating voltage; h – width of the slit of the collimator of X-ray sensor; l_1 – length of the collimation channel; l_2 – distance from the collimator to the workpiece surface; σ_x – root mean square deflections of electrons from the axis of the beam along the x axis; ε_0 – deflection of the beam from the optical axis of the electron gun; ε_m – amplitude of the beam scanning across the joint; ε_x – the position of the beam axis along the x axis.

Dependencies of odd and even harmonics from the beam deflection from the axis of the gun are shown in Fig. 2 and 3. The calculation of harmonics was produced by the formulas (13) and (14).

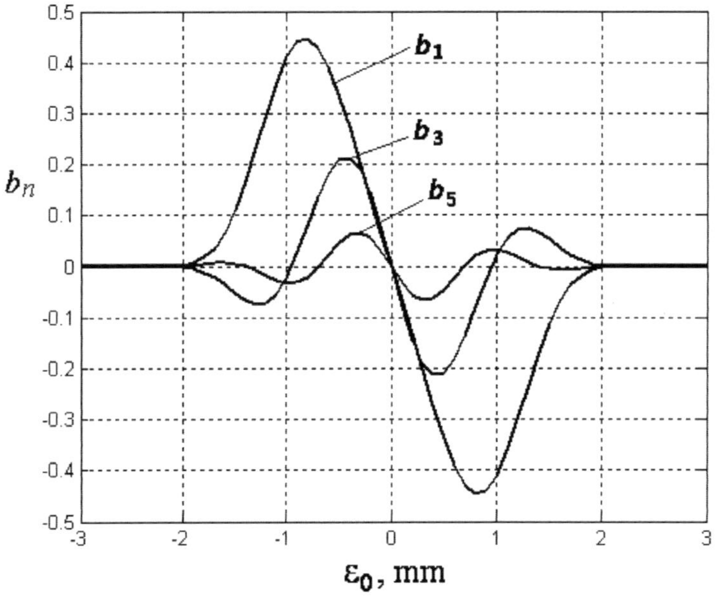

Figure 2. Dependence $b_n(\varepsilon_0)$ at ε_m=1.

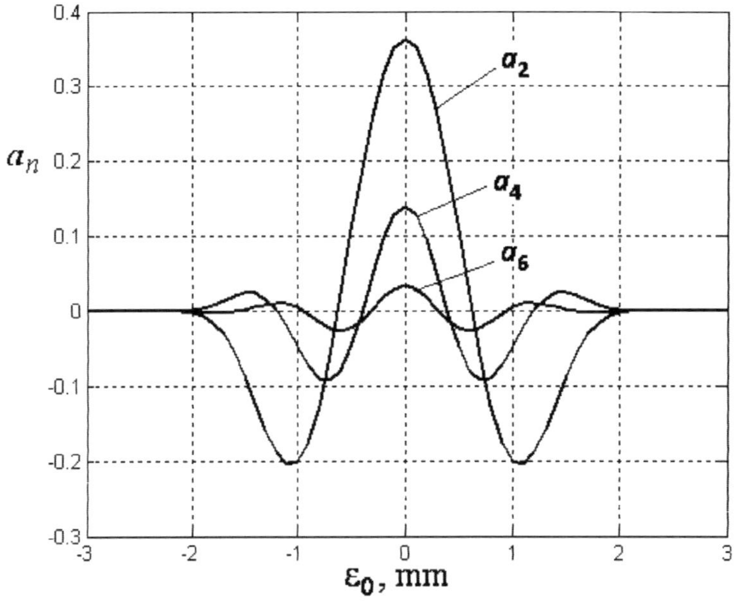

Figure 3. Dependence $a_n(\varepsilon_0)$ *at* ε_m=1.

According to the developed mathematical model of X-ray sensor the first harmonic of frequency of scanning allows to determine the beam deflection from the axis of the electron gun. The amplitude of the second harmonic is proportional to the sensitivity of the

IOP Conf. Series: Materials Science and Engineering **122** (2016) 012021 doi:10.1088/1757-899X/122/1/012021

measurement transducer. It can be used to stabilize the amplification factor of the compensation system and focusing the beam.

In welding of dissimilar materials magnetic field *B(z)* is induced by thermoelectric currents, which are distributed in welded products.

Functional block diagram of system of automatic compensation of effect of magnetic field of thermoelectric currents is shown in Fig. 4.

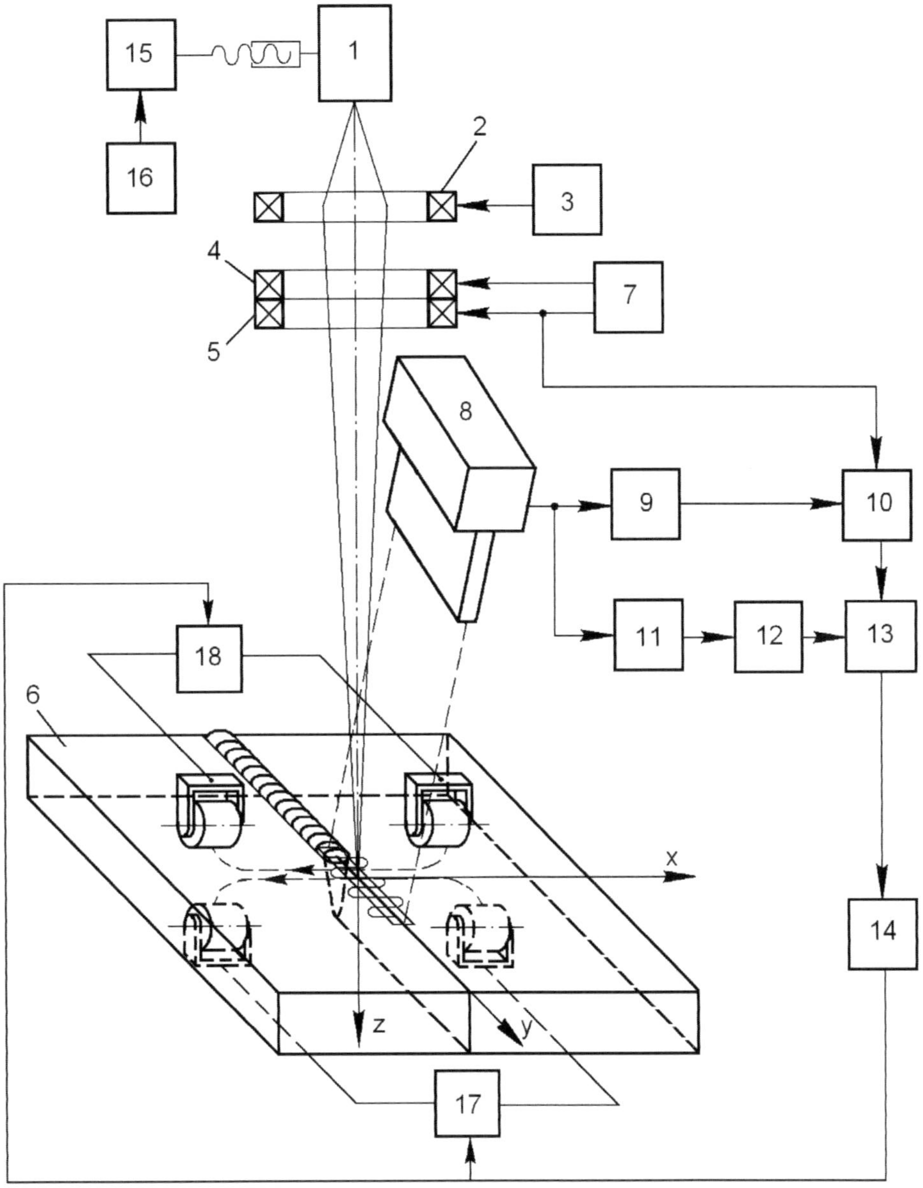

Figure 4. Functional block diagram of system of automatic compensation of effect of magnetic field of thermoelectric currents.

This system includes an electron gun 1, a focusing system 2, source of the current of focusing 3, deflecting coils 4 and 5, a generator 7 of scanning of the electron beam across the joint and along it, a block 16 aiming the electron beam at the joint, electric drive 15 moving the electron gun, the current sources 17 and 18, X-ray sensor 8, selective amplifier 9 tuned to the frequency of scanning of the electron beam across the joint, demodulator 10, the amplifier 13 with controlled amplification factor, the integrator 14, selective amplifier 11 of the second harmonic, and a rectifier 12.

Current sources 17 and 18 contain current leads which are located on the top and bottom surfaces of a welded product 6 and symmetrical to the weld in the zone of maximum temperature.

The method of compensation of the effect of magnetic fields caused by currents of thermopower includes monitoring the deflection of the electron beam from the optical axis of the electron gun by X-ray radiation from the processing zone and input into the welded product compensating currents directed oppositely thermoelectric currents. Compensating currents eliminate the beam deflection. Value of compensating currents varies depending on deflection of the electron beam from the optical axis of the electron gun.

Conclusions

The conclusions drawn are the following:

1. The considered mathematical model allows to calculate deflection of the electron beam and angle of slope of the trajectory depending on the distribution of magnetic induction in the space between the electron gun and the work piece.

2. Modeling the impact of thermoelectric currents on deflection of the beam when welding dissimilar materials agrees well with the experimental data.

3. Application of method of synchronous detection of the signal of X-ray sensor of the beam deflection allows to automate the process of compensation the impact of magnetic fields in EBW.

References

[1] Nazarenko, O. K. Avt. Svarka. 1982, No. 1, pp. 33-39.

[2] Laptenok, V. D., Murygin, A. V., Seregin, Yu. N., Braverman, V. Ya. Control of electron beam welding. Krasnoyarsk, 2000, 234 p.

[3] Nerovnyy, V. M., Konovalov, A. V., Kurkin, A. S., Makarov, E. L., Yakushin, B. F. Theory of welding processes. Moscow, 2007, 752 p.

[4] Putilov, K. A. Course of physics. Vol. 2, Moscow, 1954, 592 p.

[5] Wei, P. S., Chih-Wei Wen. Missed joint induced by thermoelectric magnetic field in electron-beam welding dissimilar metals – Experiment and scale analysis. Metallurgical and Materials Transactions B, 2002, Volume 33, Issue 5, pp 765-773.

[6] Druzhinina, A. A., Laptenok, V. D., Murygin, A. V., Seregin, Yu. N. Method for monitoring of impact of magnetic fields in the process of electron-beam welding by X-ray radiation from treatment area. Vestnik SibGAU. 2012, No. 5 (45), pp. 158-163.

IOP Conf. Series: Materials Science and Engineering **122** (2016) 012022 doi:10.1088/1757-899X/122/1/012022

System of video observation for electron beam welding process

V D Laptenok, Y N Seregin, A N Bocharov, A V Murygin, V S Tynchenko
Siberian State Aerospace University named after academician M.F. Reshetnev, 31 "KrasnoyarskiyRabochiy" prospect, Krasnoyarsk, 660037, Russia.

E-mail: sibalexbo@gmail.com

Abstract. Equipment of video observation system for electron beam welding process was developed. Construction of video observation system allows to reduce negative effects on video camera during the process of electron beam welding and get qualitative images of this process.

Methods and systems of machine vision are being actively implemented for automation of process technology and manufactures. Modern systems of machine vision allows to effectively recognize visual information. On the basis of analysis this information complicated problems of control and monitoring are solved. Very often, such systems are used where human presence is limited or not possible, or the possibility of monitoring process are limited

For example, equipment for electron beam welding (EBW) is complex, which includes electromechanical manipulators, vacuum equipment and powerful energy equipment. The high welding speed of 25 mm/s, limited visual observations pose problems to operator in control of the welding process.

Machine vision systems can be used not only for monitoring the EBW process, they also can be charged different control tasks for the process, for example, the positioning of the beam at the junction of the welded parts and the tracking of the joint during welding, runing the beam parallel to the joint or edge of the product, visual quality assessment of the welding without removing the product from the vacuum chamber, etc

However, such systems must withstand harsh conditions of electron beam welding high vacuum, the impact on the video camera thermal radiation, dust on optical system metallic vapor. And the system needs to obtain a high quality picture, avoiding exposure from the weld pool and glare from the electron beam.

Industry operating experience of television monitoring systems showed that these systems do not allow obtaining high-quality images of the welding process, because television systems cannot get rid of the exposure of weld pool, glare and degradation of image quality due to dust the lens or its protective elements. The system maintenance in industrial conditions is difficult and expensive.

Another way of monitoring EBW process [1], combining tracking of the joint, uses secondary electron emission. The electron beam scans surface of product, causing a stream of secondary electrons collected the metallic plate (collector of secondary electrons), fixed under an electron beam gun. According to the readings of this sensor forms image of the product. Further on the basis of the image, the system select the joint and brings him under the electron beam. However, the main function of this system is the pointing of the beam on the joint. Therefore, picture quality is enough to select joint, but to monitor the EBW process its not enough.

There are system video observation [2], where ordinary video camera is used as sensor. The video camera is placed directly on a movable electron beam gun, in order to adjust the optical axis of observation and the axis of the electron beam. To prevent influence of heat radiation during welding, system is equipped with water cooling. To prevent influence of metal vapor rotating glass disk is used. To obtain high-quality image used small power quartz lighter with water-cooled. However, the high level of glow of the weld pool is remain.

Consider requirements given above, system video observation for electron beam welding process, different from existing in its simplicity and accessibility for industrial exploitation, is developed [3].

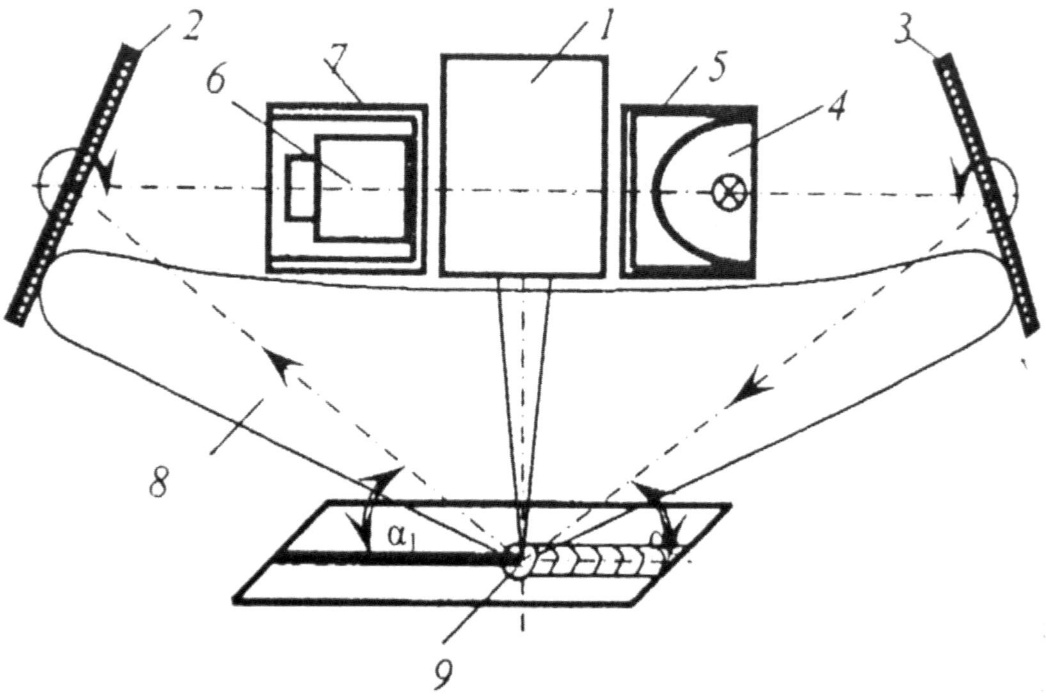

Figure 1. System of video observation for electron beam welding process:
1 – electron beam gun; 2, 3 – folding mirrors; 4 – lighter;
5 – body of lighter; 6 – video camera; 7 – thermally insulated shell; 8 – vapor cloud; 9 – welded joint

The video system (Fig. 1) consists of the video camera, the lighter and the folding mirrors. The camera and the lighter are mounted on the electron beam gun. For protection from thermal radiation and contamination of the optics, the camera placed in the thermally insulated shell. Protection against dusting of metal vapor is provided by the deviation of the optical axis of the camera and lighter from direct view of weld pool using mirrors. In view of the fact that the optimum angles of illumination and observation are provided with adjustable mirrors, placement of the protected elements is limited only by the design and dimensions of the vacuum chamber of the welding equipment. If vacuum chamber allows to place elements of a video system outside zone of expansion of vapors of welded metal, it increases the availability of the system. Also to ensure the stability of the luminous flux from the welding zone it is advisable to use a mirror with a mirror layer of welded material that can be produced by any suitable known method. Since the brightness of the radiation of the weld pool is high, the system uses a concentrated light source, avoiding the exposure of the camera from glare and radiation of the weld pool.

The system of video observation can be assembled on the basis of the video cameras produced by companies specializing in video surveillance and security systems. However, there are a number of conditions that need to be considered at choosing a camera. The video camera's resolution should be not less than 520 TVL for analog cameras and not less than 720×540 for digital. The minimum sensitivity of video camera should be 0.01 to 1 Lux. The video camera should be suitable for effective observation day and night. The video camera must be equipped with an auto-iris or have a block to control the aperture. The auto-iris is used for observations of objects with variable illumination, which is suited to the observation of weld pool, brightness of which changes rapidly during the welding process. The distance from the electron beam gun to the welded parts average from 80 to 200 mm, therefore, to monitor the welding process ideal camera mast have dynamic focal length. The exploitation of the video system revealed that the increase in 4 – 5 times is enough for observation of the welding process. And only in rare cases, the zoom lenses change focal length more than six times are needed. You should also consider that system of video observation must operate in a vacuum, so the choice of video camera is fulfilled with the absence of many elements that are critical to low pressure.

The appearance of the system of video observation shown in Fig. 2

Figure 2. The appearance of the system of video observation for electron beam welding process

Industrial designs the system of video observation for electron beam welding process are implemented in production and successfully used by JSC " Krasmash."

References

[1] Nazarenko O. K., Shapoval V.I., Loskutov G.A 1993 Observation of the process of electron-beam welding and automatic tracking of joint *Automatic welding* № 5. - p. 35-38.

[2] Haynie, T. J. 1990 A new video viewing system for moving in-vacuum EB guns *Abstr. Pap. Present. 71st AWS Annu. Meet. and 21st Int. AWS Braz. and Solder. Conf., Miami, Fla, Apr. 22-27* p. 202-203.

[3] Kolmykov V.A., Uspenskiy N.V., Laptenok V.D., Seregin Y.N., Bocharov A.N., Kotelnikova S.V. 2010 Method for protection of optical systems from deposition by electron-beam welding Pat. № 2391190 Bul. № 16.

IOP Conf. Series: Materials Science and Engineering **122** (2016) 012023 doi:10.1088/1757-899X/122/1/012023

The structural defects formation in ZnS under electron irradiation with energy 400 KeV

Y Y Loginov, I V Kovalev, P V Zelenkov
Siberian State Aerospace University
31 Krasnoyarskiy Rabochiy prospect, Krasnoyarsk, 660037, Russia.

E-mail: loginov@sibsau.ru

Abstract. The formation of structural defects in grown from the vapour phase ZnS crystals after electron irradiation with energy of 400 keV and intensity of $1 - 4 \cdot 10^{19}$ e/cm^2·s in situ in electron microscope was studied. It is shown that under electron irradiation the small dislocation loops in size of 2.5 - 45 nm and a density of $1.4 \cdot 10^{11}$ cm^{-2}, as well as voids and fine particles of a new phase in size ≤10 nm are formed. These features can be identified from an analysis of moiré fringe contrast as phase of ZnO_2.

1. Introduction

The study of the formation of structural defects and changes in the properties of irradiated semiconductors is of interest both in terms of increasing their radiation resistance, and ability to manage the properties of materials under irradiation [1-5]. This is especially important for semiconductor compounds A_2B_6, because of their high sensitivity to radiation and insufficient study of these issues.

Zinc sulfide, belonging to the group of compounds A_2B_6, is used for the detection of charged particles. The high efficiency of radiative recombination of electrons and holes of the material makes it possible to use it as an active element of the semiconductor laser. Further, ZnS is a promising material of the power quantum optics.

The formation of structural defects in semiconductors during electron irradiation is possible in the conditions of electron irradiation with a high intensity ~10^{17}-10^{19} e/cm^2·s that is achievable, for example, in the transmission electron microscope (TEM). These intensities several orders of magnitude higher than the intensity of radiation in conventional industrial accelerators (~10^{12}-10^{14} e/cm^2·s), which means that in TEM on the area occupied by one atom drops ~ 10^2 - 10^4 electrons per second.

The formation of structural defects in materials irradiated by electrons with energies above the threshold energy of radiation-induced defects formation (E_T) under irradiation *in situ* in a high-voltage electron microscope (E ≥ 1 MeV) has been well studied for Si, Ge [3,6] and metals [7-9]. The formation of structural defects in ZnS under electron irradiation with above-threshold energies ($E > E_T$) has been insufficiently studied.

The purpose of work was to study the influence of electrons with an energy of 400 keV ($E > E_T$) on the formation of structural defects in ZnS.

2. Experimental

Crystals of ZnS with resistivity $\rho \approx 7{\cdot}10^3$ $\Omega{\cdot}$cm were prepared using a modified Piper-Polich technique. The crystals were synthesized directly from the vapour phase at a temperature of 1500 ^{0}C using polycrystalline starting material, under an argon ambient.

Bulk ZnS crystals were prepared in thin foil form by sequential mechanical polishing and argon ion milling at 5kV, 20μA and 15^0 with liquid nitrogen cooling until electron transparent. Improved sample foils were obtained following iodine reactive ion sputtering for 5 to 10 min at 3kV, 10μA and 15^0 at room temperature. To completely remove the surface defects formed by ion milling as appropriate thin foil was subjected to chemical polishing for 2-3 seconds in a dilute solution of HPC (a supersaturated solution of CrO_3 in H_3PO_4 heated to 60 ^{0}C (1 part), mixed with concentrated HCl (2 parts)).

Samples were examined and irradiated in a JEOL 4000EX-II electron microscope operated at 400keV ($E > E_T$) and $j = 1 - 4{\cdot}10^{19}$ e/cm^{2}s.

3. Results and Discussion

The electron microscopic study of ZnS crystals before irradiation show that they contain of well developed native dislocations and long stacking faults. Also well-developed parallel strips (striations), indicating polytypic modifications of ZnS, are also observed in all cases. This is especially true when the diffraction vector **g** is perpendicular to these strips.

Irradiation of ZnS by electrons with an energy of 400 eV in TEM accompanied by the formation of small dislocation loops. Fig. 1 shows the formation of dislocation loops in ZnS after electron irradiation for 5 minutes. Dislocation loops have a size of 2.5 - 45 nm at a density of $1.4{\cdot}10^{11}$ cm^{-2}. In addition, striations due to the polytype structure of this compound are more readily apparent. The ZnS foil shown in Fig.1 was given a brief chemical polish following Ar^+ and I^+ milling to remove the remnant artefact structure.

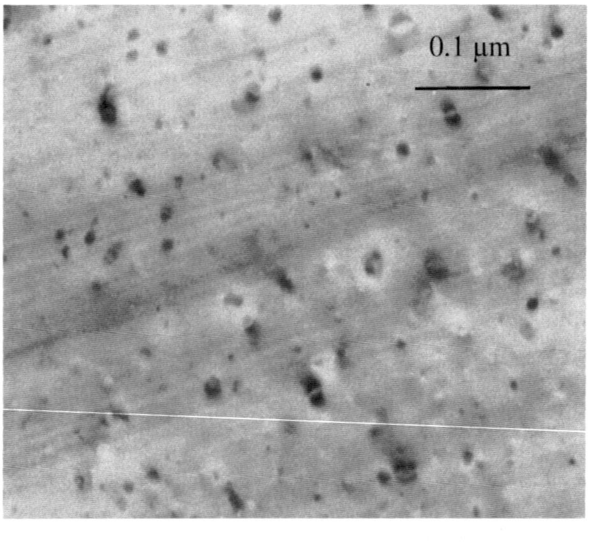

Figure1. TEM image of structural defects in the ZnS irradiated with 400 keV electron beam of intensity $1{\cdot}10^{19}$ e/cm^{2}s for 5 minutes.

The image with a high resolution (Fig. 2) shows on atomic-scale the formation of dislocation loops in ZnS after electron irradiation. In the image processing has been applied at least 20 diffracted beams. Extra-planes of dislocations are marked by arrows in Fig. 2. Burgers contours are also drawn. From Fig. 2 follow that the plane of dislocation loop location is the plane {111}, and the Burgers vector of the boundary dislocations is **b** = a/6<112>.

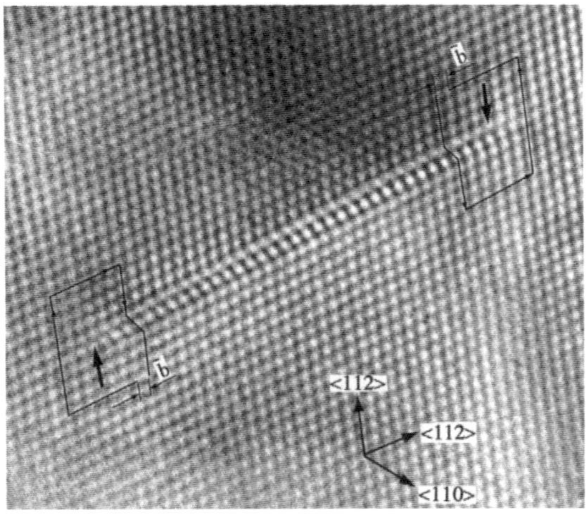

Figure 2. HREM image of 400 keV electron beam irradiated ZnS with intensity $4 \cdot 10^{19}$ e/cm^{2}s for 5 minutes

Subsequent electron beam irradiation of ZnS caused growth, transformation, shrinkage and movement of dislocation loops, the formation of new loops, and then the formation of voids. Fig. 3a shows a high resolution plan-view image of dislocation loop formed in ZnS after irradiation for 5 min. In Figure 3b shows the same place after irradiation for 15 minutes. It is seen the formation of voids, which are formed by the accumulation of vacancies, indicating that further decomposition of the material under electron irradiation. At the same time, the crystal structure of the material is preserved, as indicated by electron diffraction pattern obtained from the same place of sample and shown in the upper right corner of Fig. 3b.

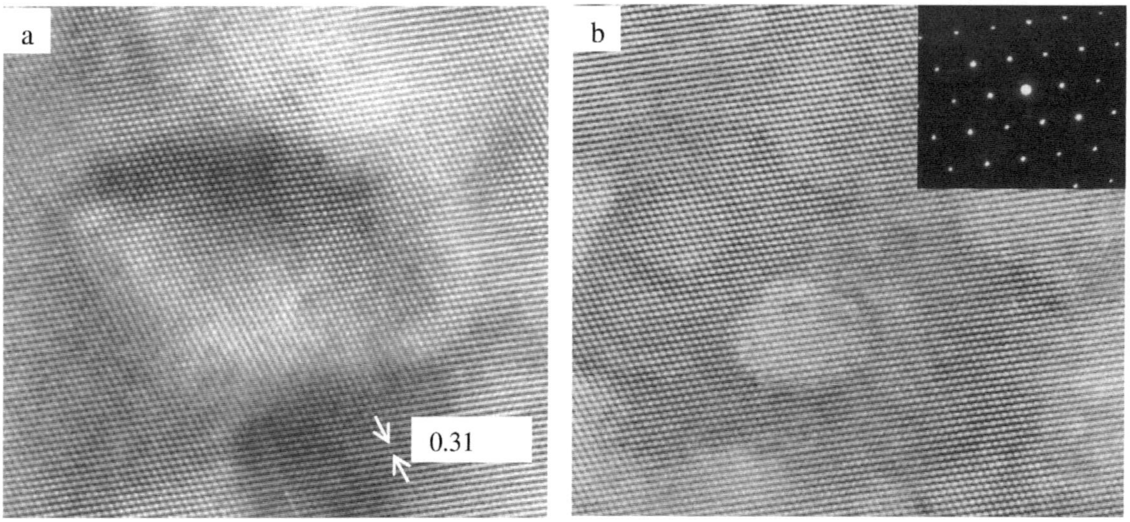

Figure 3. <110> HREM image of 400 keV electron beam irradiated ZnS (the same place of thin foil) with intensity $4 \cdot 10^{19}$ e/cm^{2}s for 5 minutes (a) and 15 minutes (b)

Formations of fine particles size ≤ 10 nm with specific moiré fringe contrast were also observed. Creation of modern semiconductor devices of ultra-sizes requires control not only the presence and density of defects such as dislocations, dislocation loops, stacking faults, and microtwins, but also control defects of size of ≤ 10 nm. As shown earlier, the formation of new phases in such sizes in materials A_2B_6 possibly under the influence of electrons and ions [10, 11].

Typically, microanalysis of the particles of this size is difficult to technically implement. However, the use of technique moiré contrast, formed from the particles in the high-resolution images, allowing, in certain cases, to determine the chemical composition of these particles. In this case, the period of the moiré contrast D, occurring when rotating the lattices with the parameters d_1 and d_2 are at an angle α, can be determined according to [12].

As a result of analysis of fine particles formed in ZnS crystals, irradiated by electrons with energy of 400 keV, was found that the small particles with a moiré fringe contrast can be identified as the phase of ZnO_2, which is consistent with previous results [1, 10]. So, the decomposition products produced by 400keV electron beam irradiated ZnS is a ZnO_2. The tendency for the initial formation of oxides of Zn indicates the preferential removal of the anion leading to accelerated oxidation [13].

4. Summary

It was found that the irradiation of ZnS crystals by electrons with energy of 400 keV is accompanied by the formation of dislocation loops, and further shrinkage and movement of loops, the formation of voids and fine particles of the new phase, which can be identified from an analysis of a moiré fringe contrast as ZnO_2. Regularities of structural defects formation in ZnS can be used to solve problems of management type, density and spatial distribution of defects in the crystal structure, which is important for the implementation of the limiting parameters of microelectronic devices.

References.

[1] Loginov Y, Brown P, Durose K *The Regularities of Structural Defect Formation in Semiconductors II–VI* (Logos, Moscow, 2003)
[2] Kozlovskii V, Lebedev A, Bogdanova E, Seredova N 2015 *Phys. Techn. Semicond.*, **49** 1198
[3] Aseev A, Fedina L, Hoehl D, Bartsch H *Clusters of interstitial atoms in silicon and germanium* (Akad. Verlag, Berlin, 1994)
[4] Grushka O, Maslyuk V, Chupyra S, Mysliuk O 2012 *Phys. Techn. Semicond.*, **46** 327
[5] Lavagne S, Levade C, Vanderschaeve G 2006 *Mater. Science and Engineering: B*, **128** 1
[6] Hirata M and Kiritani M 1983 *Phys.B.*, **116** 616
[7] Nakai K, Kinoshita C, Muroo Y, Katajima S 1983 *Phil. Mag. A.*, **48** 215
[8] Urban K 1979 *Phys. Stat. Sol.(a)*, **56** 157
[9] Lazarenko V, Platov Y, Pletnev M 1980 *Phys. Metals and Metallography*, **50** 164
[10] Loginov Y, Brown P, Humphreys C 1995 *Microsc. Semicond. Mater.- Bristol, London: Inst. Phys. Conf.*, **146** 431
[11] Loginov Yu and Brown P 1992 *Phys. Stat. Sol. (a)*, **132** 323
[12] Hirsch P, Howie A, Nicholson R, Pashley D and Whelan M *Electron microscopy of thin crystals* (Mir, Moscow, 1967)
[13] Thangaraj N and Wessels B 1990 *J. Appl. Phys.*, **67** 1535

Global optimization on set of mixed variables: continuous and discrete with unordered possible values

A S Mikhalev[1], A I Rouban[2]

Computer Science Dept, Institute of Space and Information Technologies, Siberian Federal University, Street of academician Kirenskogo, nimber 26, Krasnoyarsk region, Krasnoyarsk 660074, Russia

E-mail: [1] spaming1@yandex.ru; [2] ai-rouban@mail.ru

Abstract. The algorithms of global non-differentiable minimization of functions on set of the mixed variables: continuous and discrete with unordered specific possible values are constructed. The method of optimization is based on selective averaging of required variables, on adaptive reorganization of the sizes of admissible domain of trial movements and on use of relative values for minimised functions. Existence of discrete variables leads to solution of a sequence of global minimization problems of the functions in space of only continuous variables at the presence: 1) of their inequality restrictions for each problem; 2) of the general inequality restrictions for all problems (i.e. at the absence of dependence of functions fore inequality restrictions from discrete variables). In the first case, presence of discrete variables with unordered non-numeric possible values leads to solution of sequence of problems of global minimization of multiextreme functions on set only of continuous variables at the presence of their inequality restrictions. As a result, among the received optimum solutions the best is selected. In the second variant all minimized functions is convoluted in each sampling point in one multiextreme function and this function is minimised on continuous variables.

1. Introduction

The problem of search of a global extremum [1-7] is very difficult. Its specificity are caused by possible breaks in many-extreme optimised functions and in functions of restrictions, the presence of noises, the limited set of possible values of the required variables and also the presence of discrete variables in addition to continuous variables.

The area of the mixed global optimization is applied for solution of optimization problems in different scopes. Among them are process industry and the financial engineering, engineering design, chemical engineering, the theory of control and process operations research, automobile and aircraft manufacturing and many others. Application of algorithms of mixed global optimization approach allows to solution problems of calculation of optimal parameters of the equipment, optimal conditions of functioning of process, optimal design of networks, building of schemes, etc.

Methods of solution of problems of search of optimal value in space continuous and discrete variables are discussed in the literature since the early 1980's. Approaches to their solution can be mainly separated into two classes. The first class – is heuristic methods, which don't guarantee that when the search is complete, the found solution will be optimal. Other class – deterministic methods – in turn, stop search with the guaranteed solution or indicate, that the problem has no integer solution.

The deterministic methods for solution of this class problem usually are based on the successive solutions of closely related problems of nonlinear programming. The basic concept which is underlies the algorithms of solution of problems – generation and refinement of estimates of optimal solution. Algorithms differ on way of forming of these estimates and on type of sequence of subtasks, which are solved for receiving these estimates [8-11].

Example of the heuristic approach to solution of problems is the method, which combined together a global search phase and a local search phase [12].

In this paper on the basis of a method [13–15] the algorithms of the decision of two problems of global optimisation with the mixed variables are received.

Discrete variables, based on properties of their possible values, conditionally can be divided into 2 classes: with **unordered** and **ordered** possible values. In turn, possible values of discrete variables of both classes can accept **non-numerical** and **numerical** possible values.

In this paper discrete variables with disorder non-numerical possible values are considered, and for each of them there is a minimized function and the inequality restrictions. Restrictions can be and the general for all minimized functions.

In the presence of discrete variables with the ordered possible values in other papers the approach based on one-for-one transition from discrete variable to a continuous variable will be considered. To each possible value of a discrete variable the continuous interval of values is put in conformity. Transition from continuous values to the discrete is carried out both at formation of trial movements and at calculation of required variables on each working step. This way allows to apply a method of global optimization in space of continuous variables to optimization on set mixed variables.

2. Statement of problem

The problem of search of a conditional global minimum of objective function of many variables $f(x, y)$ on set of the mixed variables: continuous and discrete with unordered specific possible values at the presence of inequality restrictions is solved:

$$f(x, y) = \min, \quad \varphi_j(x, y) \leq 0, \; j = \overline{1, m(y)}. \tag{1}$$

Here: $x = (x_1, ..., x_k)$ – vector k of continuous variables, y – discrete variable with r unordered non-numerical possible values $y_1, \ldots, y_r$.

The presence of additional discrete variable with unordered possible values leads to solution of r independent problems of global minimization (1) only on continuous variables at the presence of inequality restrictions:

$$\left[f_\mu(x) = \min, \; \varphi_{j\mu}(x) \leq 0, \; j = \overline{1, m_\mu}\right], \; \mu = \overline{1, r}. \tag{2}$$

Here:

$$f(x, y_\mu) \equiv f_\mu(x), \quad \varphi_j(x, y_\mu) \equiv \varphi_{j\mu}(x).$$

As a result, among the obtained optimum solutions $\overset{*}{x}_\mu$, $\mu = \overline{1, r}$ the best solution $\overset{*}{x}_{\mu^*}$ is selected. It corresponds to the optimum value $\overset{*}{y}_{\mu^*}$ (with number μ^*) of a discrete variable.

3. The method selective averaging of required variables

We provide the solution of a r problems of global optimization (2) by using basic algorithm of method with selective averaging of required continuous variables [13-15]. The method is based on separation in time of trial and working steps, uniform distribution of sampling points in admissible region, selective averaging of required variables by results of the experimental data, obtained in sampling points, adaptive reorganization at each working step of the sizes of rectangular domain of trial movements. Further we will describe algorithm of finding of a global minimum (2) at the fixed μ.

For receipt of sampling points $x_{v,\mu}^{(i)}$, $i = \overline{1, n}$ in rectangular domain Π_μ^l with a center in a point $x_{v,\mu}^l$ is sequentially generated uniform distribution of points:

$$x_{v,\mu}^{(i)} = x_{v,\mu}^l + \Delta x_{v,\mu}^l u_v^{(i)}, \; u_v \in [-1; 1], \; v = \overline{1, k}, \; i = 1, 2, \ldots \tag{3}$$

and from them points, which satisfy to inequalities restrictions, are left. These points are in admissible region $X_\mu^l = \Pi_\mu^l \cap X_\mu$.

The starting point x_μ^0 and sizes $\Delta x_{v,\mu}^0$ of rectangular domain Π_μ^0 of trial movements are selected so, that Π_μ^0 covered the admissible region X_μ or the part where the required global minimum is located.

In sampling points the minimized functions $f_\mu^{(i)} \equiv f_\mu(x^{(i)}), i = \overline{1,n}$ are calculated. After, the position of minimum is specified.

At movement to the extremum occurs adaptation (most of reduction) of the region of search movements. New value $x_{v,\mu}^{l+1}$, on average closer to position of a global minimum of the optimized function f_μ, and the sizes of rectangular domain of search movements $\Delta x_{v,\mu}^{l+1}$, $v = \overline{1,k}$, are calculated on formulas [14]:

$$x_{v,\mu}^{l+1} = x_{v,\mu}^{l} + \Delta x_{v,\mu}^{l} u_{v,\mu,\min},\ u_{v,\mu,\min} = \sum_{i=1}^{n} u_{v,\mu}^{(i)} \bar{p}_{s,\min}^{(i)},\ v = \overline{1,k}, \tag{4}$$

$$\bar{p}_{s,\min}^{(i)} = \frac{p_s(g_{\mu,\min}^{(i)})}{\sum_{j=1}^{n} p_s(g_{\mu,\min}^{(j)})},\ g_{\mu,\min}^{(i)} = \frac{f_\mu^{(i)} - \widehat{f}_{\mu,\min}}{\widehat{f}_{\mu,\max} - \widehat{f}_{\mu,\min}},$$

$$\Delta x_{v,\mu}^{l+1} = \gamma_q \Delta x_{v,\mu}^{l} \Big(\sum_{i=1}^{n} | u_{v,\mu}^{(i)} |^q\ \bar{p}_{s,\min}^{(i)}\Big)^{1/q},\ v = \overline{1,k},$$

$$l = 0, 1, 2, ...; [0 < \gamma_q, q \in \{1, 2, ...\}, 0 < s].$$

Here $\widehat{f}_{\mu,\max} = \max\{ f_\mu^{(i)}, i = \overline{1,n}\}$, $\widehat{f}_{\mu,\min} = \min\{ f_\mu^{(i)}, i = \overline{1,n}\}$, $p_s(\cdot)$ – kernel, s – a degree of selectivity (ore selectivity coefficient) of kernel $p_s(\cdot)$. The kernels are normalized on 1 on a system of n sampling points: $\sum_{i=1}^{n} \bar{p}_{s,\min}^{(i)} = 1$. The argument of kernel – is non-dimensional variable, it always lies in interval $[0; 1]$: $0 \le g_{\mu,\min}^{(i)} \le 1$.

When approaching to minimum the region of trial movement reduced, and thus, there is more exact tracking of position of extremum. The criterion of stop of search process is the condition of reduction (at some l) of size of region of variation of variables to the set value:

$$\max\left\{ \frac{\Delta x_{v,\mu}^{l}}{\Delta x_{v,\mu}^{0}}, v = \overline{1,k} \right\} \le \varepsilon \text{ or } \frac{\Delta x_{v,\mu}^{l}}{\Delta x_{v,\mu}^{0}} \le \varepsilon, v = \overline{1,k}. \tag{5}$$

After finding of global minimum for each of r functions $\{x_\mu^*,\ f_\mu(x_\mu^*),\ \mu = \overline{1,r}\}$ the best solution $x_{\mu^*}^*$, $f_{\mu^*}(x_{\mu^*}^*)$ is selected:

$$\min\{ f_\mu(x_\mu^*), \mu = \overline{1,r}\} = f_{\mu^*}(x_{\mu^*}^*). \tag{6}$$

Optimal value of discrete variable has number μ^*.

4. Example 1

We form 4 thirteen-extreme functions:

$\mu = 1$: $f_1(x_1, x_2) = \min\{4 | x_1 - 2 | + 5 | x_2 - 2 | + 16; 6 | x_1 - 2 |^{1.1} + 7 | x_2 - 6 |^{0.7} + 4;$

$3 | x_1 - 2 |^{1.3} + 3 | x_2 - 10 |^{1.3} + 10; 7 | x_1 - 6|^{0.8} + 7 | x_2 - 10 |^{0.6} + 2;$

$5|x_1-10|^{1.8}+5|x_2-10|^{1.8}+14; 8|x_1-10|^{0.9}+8|x_2-6|^{1.1}+6;$
$6|x_1-10|^{0.5}+7|x_2-2|^{0.5}+12; 2|x_1-6|^{1.8}+5|x_2-2|^{1.5}+8;$
$6|x_1-4|^{0.9}+5|x_2-4|^{0.7}+9; 4|x_1-4|^{1.2}+4|x_2-8|^{1.5}+3;$
$6|x_1-8|^{1.6}+5|x_2-8|^{0.6}+7; 5|x_1-8|^{1.3}+7|x_2-4|^{0.9}+5; 5|x_1-6|^{2}+5|x_2-6|^{2}\};$
$\mu=2: f_2(x_1,x_2)=\min\{3|x_1+2|^{1.1}+3|x_2-2|^{1.3}+9; 7|x_1+2|^{0.9}+6|x_2-6|^{0.7}+4;$
$3|x_1+2|^{0.9}+4|x_2-10|^{0.3}+7; 6|x_1+6|^{0.9}+6|x_2-10|^{0.3}+9;$
$5|x_1+10|^{0.7}+3|x_2-10|^{1.6}+5; 3|x_1+10|^{0.8}+5|x_2-6|^{1.7}+6;$
$4|x_1+10|^{0.9}+2|x_2-2|^{1.1}+11; 2|x_1+6|^{1.8}+3|x_2-2|^{1.8}+8;$
$6|x_1+4|^{1.5}+5|x_2-4|^{1.5}+19; 4|x_1+4|^{1.3}+4|x_2-8|^{1.3}+13;$
$3|x_1+8|^{0.9}+4|x_2-8|^{0.9}+15; 4|x_1+8|^{0.6}+7|x_2-4|^{0.6}+17; 6|x_1+6|^{0.8}+6|x_2-6|^{0.6}+2\};$
$\mu=3: f_3(x_1,x_2)=\min\{4|x_1+2|^{2.4}+5|x_2+2|^{1.6}+7; 3|x_1+2|^{1.8}+2|x_2+6|^{2.2}+14;$
$3|x_1+2|^{1.7}+3|x_2+10|^{1.9}+9; 4|x_1+6|^{1.4}+2|x_2+10|^{1.8}+10;$
$6|x_1+10|^{0.8}+7|x_2+10|^{1.5}+13; 5|x_1+10|^{1.3}+3|x_2+6|^{1.4}+8;$
$4|x_1+10|^{1.6}+5|x_2+2|^{1.3}+11; 2|x_1+6|^{1.7}+5|x_2+2|^{1.9}+12;$
$6|x_1+4|^{0.4}+5|x_2+4|^{1.6}+18; 4|x_1+4|^{1.8}+4|x_2+8|^{0.9}+20;$
$6|x_1+8|^{1.6}+5|x_2+8|^{2.4}+22; 5|x_1+8|^{1.5}+7|x_2+4|^{0.6}+16; 7|x_1+6|^{1.1}+6|x_2+6|^{0.9}+4\};$
$\mu=4: f_4(x_1,x_2)=\min\{4|x_1-2|^{1.5}+5|x_2+2|^{1.9}+11; 6|x_1-2|^{2.2}+7|x_2+6|^{2.2}+17;$
$3|x_1-2|^{1.2}+3|x_2+10|^{0.8}+9; 7|x_1-6|^{1.7}+7|x_2+10|^{1.7}+19;$
$5|x_1-10|^{0.6}+5|x_2+10|^{1.5}+13; 4|x_1-10|^{2.4}+5|x_2+6|^{2.4}+23;$
$6|x_1-10|^{1.1}+7|x_2+2|^{1.6}+15; 2|x_1-6|^{1.5}+5|x_2+2|^{1.5}+21;$
$2|x_1-4|^{2.5}+3|x_2+4|^{1.5}+18; 3|x_1-4|^{1.3}+2|x_2+8|^{2.3}+10;$
$5|x_1-8|^{1.7}+4|x_2-8|^{0.7}+14; 3|x_1-8|^{2.1}+4|x_2+4|^{1.1}+22; 5|x_1-6|^{1.1}+5|x_2+6|^{1.3}+6\}.$

Each of them corresponds to value of discrete variable. The required global minimum is in point $\mu^*=1: x^*=(6;\,6)$, и $f_1(x^*)\equiv f(x^*,y_1^*)=0$.

Each of functions has the inequality restrictions, which allocate the admissible region in circle and on its border:

$$\mu=1: (x_1-6)^2+(x_2-6)^2-25\le 0;\quad \mu=2: (x_1+6)^2+(x_2-6)^2-25\le 0;$$
$$\mu=3: (x_1+6)^2+(x_2+6)^2-25\le 0;\quad \mu=4: (x_1-6)^2+(x_2+6)^2-25\le 0.$$

For $\mu^*=1$ spatial form of the minimized function, the lines of equal levels for it, restriction and first steps of movement to global minimum are presented on Figure 1. The conditional global extremum is in point: $\mu^*=1: x^*=(6;\,6)$.

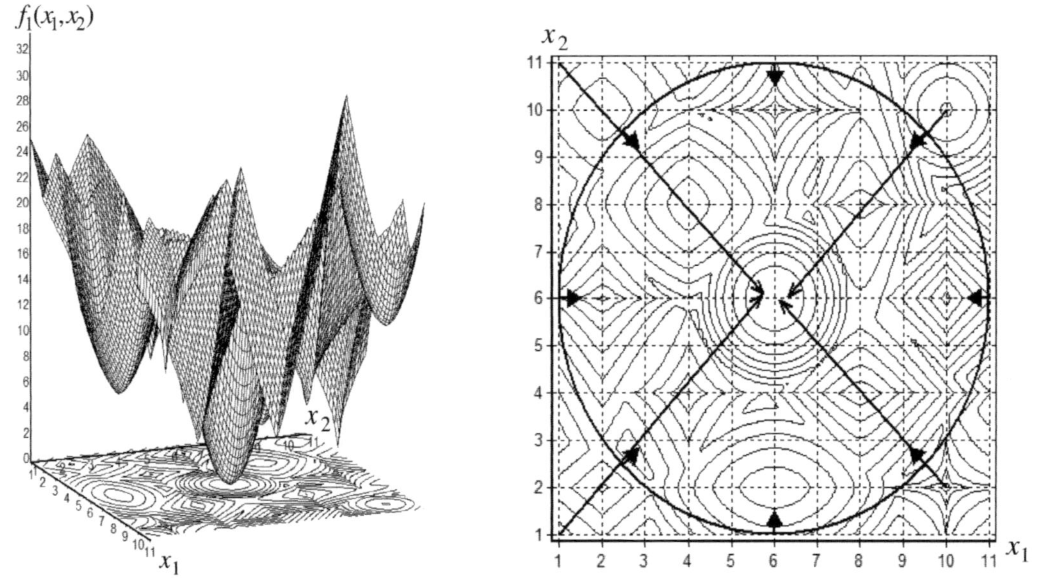

Figure 1. A perspective view of the first function (for $\mu = 1$), equipotential lines, permissible area and first steps of algorithm

Let's show operability of the offered algorithm. The complex characteristic of process of search of global minimum is the estimate $\hat{P}_{\text{correct}}$ of probability of hitting of the obtained required variables in ε neighborhood of the true solution. The received solution x^* gets in ε neighborhood of true position of global minimum x^{**} , if:

$$\frac{|x_\nu^* - x_\nu^{**}|}{\Delta x_\nu^0} \le \varepsilon, \ \nu = \overline{1,k}\,, \tag{7}$$

i.e. if the modules of relative misalignments for all coordinates don't output abroad ε.
Here: Δx_ν^0 – components which determine the sizes of initial rectangular domain of trial movements.

Let's consider dependence $\hat{P}_{\text{correct}}$ from number of sampling points n. For this purpose N realization of process of search of minimum carried out. The estimate of probability $\hat{P}_{\text{correct}}$ will be equal to the relative frequency of hitting of the obtained solutions $x_j^*, j = \overline{1,N}$ in rectangular neighborhood of true solution.

The parameters of the algorithm of minimization the following: $x^0 = (0;0)$, $\Delta x^0 = (10;10)$, $\gamma_q = 1$, $q = 2$, kernel on the minimized function parabolic with degree of selectivity $s = 100$. Research parameters: number of realization $N = 101$, size of neighborhood of true solution $\varepsilon = 0.0005$. The received dependence $\hat{P}_{\text{correct}}$ (by $\mu = 1$) from number of sampling points n is presented on Figure 2.

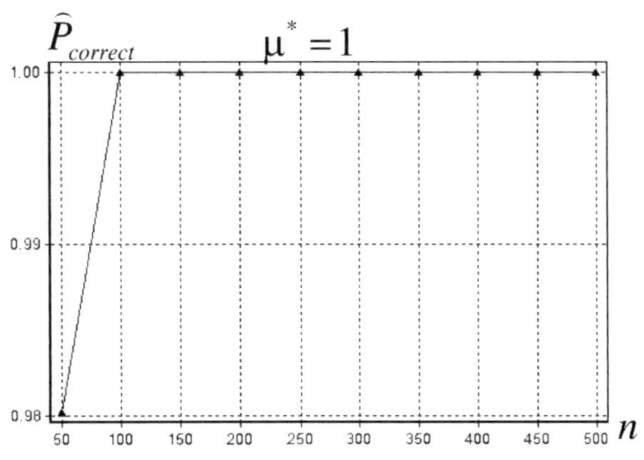

Figure 2. Dependence of estimate of probability of hitting to neighborhood of the true solution from a number of sampling point n

The estimate of probability of hitting of the obtained solution in neighborhood of the true solution is equal to 1 by $n \geq 100$.

5. Statement of problem 2

The problem of search of a conditional global minimum of objective function of many variables $f(x, y)$ with unordered non-numerical possible values at the same inequality restrictions, i.e. at functions of restrictions, which not depend on discrete variable $[\varphi_j(x) \leq 0,\ j = \overline{1, m}]$ is solved:

$$f(x, y) = \min,\ \ \varphi_j(x) \leq 0,\ j = \overline{1, m}. \tag{8}$$

We pass from solution of problems of global optimization (3) to solution of one problem:

$$f(x) = \min,\ \ \varphi_j(x) \leq 0,\ \ j = \overline{1, m}. \tag{9}$$

Here in each fixed point x the function

$$f(x) = \min\{f_1(x), f_2(x), ..., f_r(x)\}. \tag{10}$$

After finding of position x^* of global minimum of function $f(x)$ we define number μ^* of optimal value $y^*_{\mu^*}$ of discrete variable y:

$$\min\{f_1(x^*), f_2(x^*), ..., f_r(x^*)\} = f_{\mu^*}(x^*). \tag{11}$$

Computing costs at decision reception are reduced almost by r times.

6. Example 2

Let's consider the same problem, as in an example 1 but we set single admissible region inside and on border of circle $x_1^2 + x_2^2 - 121 \leq 0$ radius 11 with the center in origin of coordinates. A perspective view of the minimized function $f(x_1, x_2)$ (look (10)), the lines of equal levels, restriction and first steps of movement to global minimum are presented on Figure 3.

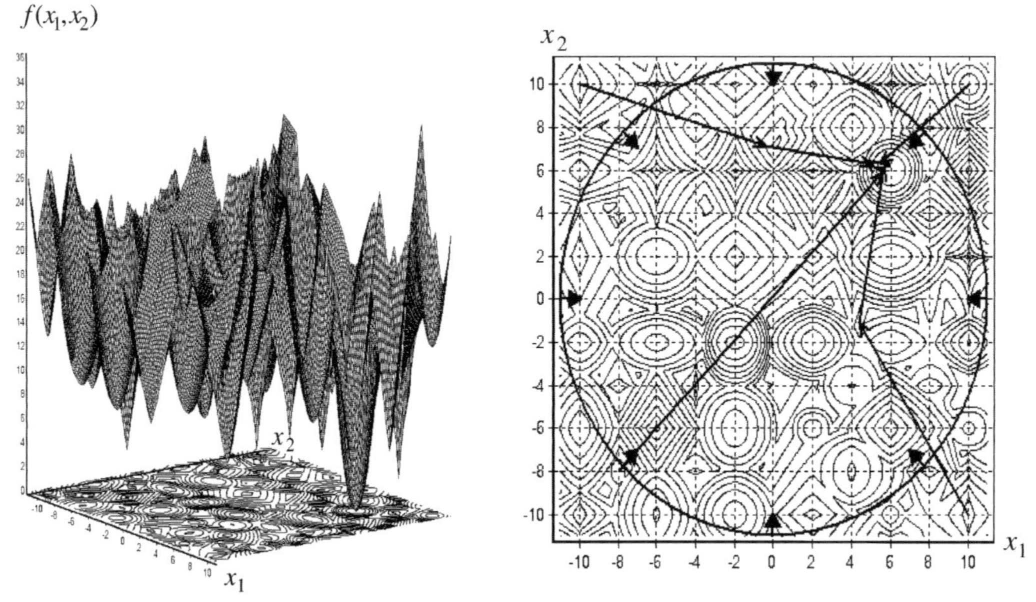

Figure 3. A perspective view of function $f(x_1, x_2)$, equipotential lines, permissible area and first steps of algorithm

Using the scheme, described in section 6, we investigate dependence of estimate $\widehat{P}_{\text{correct}}$ of probability of hitting of the obtained required variables in ε neighborhood of the true solution from a number n of sampling point at $50 \le n \le 500$ and $\varepsilon = 0.01$.

The parameters of the algorithm of minimization the following: $x^0 = (0; 0)$, $\Delta x^0 = (10; 10)$, $\gamma_q = 1$, $q = 2$, kernel on the minimized function parabolic with degree of selectivity $s = 100$. The number of realization $N = 101$. The received dependence $\widehat{P}_{\text{correct}}$ (by $\mu = 1$) from number of sampling points n is presented on Figure 4.

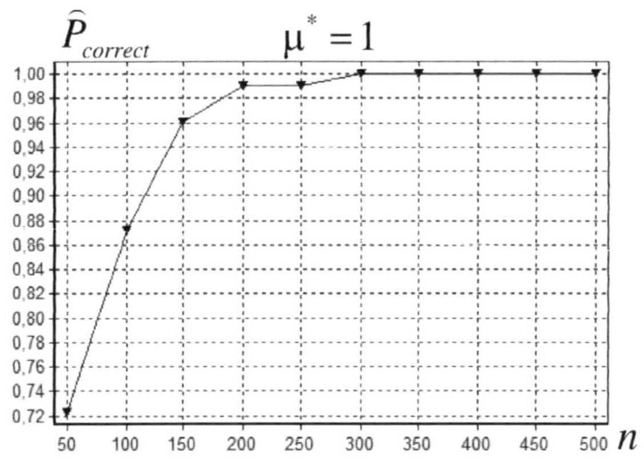

Figure 4. Dependence of estimate of probability of hitting to neighborhood of the optimum true from a number of sampling point n

Apparently from graph on Figure 4 with increase of volume of sampling the estimate of probability of hitting of the obtained solution in neighborhood of the true solution increases and reaches 1 при $n \geq 300$.

7. Conclusion

The offered approach to synthesize algorithms of global optimization at the mixed required variables can be applied to solution: 1) of problems of multicriteria global optimization, 2) of systems of nonlinear equations, 3) of searching of principal minima.

The developed method of global optimization allows to count simply enough on each working step a site of a global extremum. A method kernel is selective averaging of required variables. At the expense of such averaging is carried out more exact movement to an extremum. Transition in algorithms to dimensionless (relative) variables essentially increases speed of convergence, simplifies structure of algorithms and accordingly reduces quantity of a tuned parameters in them. All it is especially important at the account of a considerable quantity of heterogeneous restrictions and at the decision of problems of multicriterion global optimization.

References

[1] Evtushenko Yu G 1971 Numerical methods for finding global extrema (case of a non-uniform mesh) *J. Computational Mathematics and Mathematical Physics* **11:6** 38-64

[2] Strongin R G 1978 *Numerical methods in mulextreme problems* (Moscow: Publishing House «Nauka») p 239

[3] Vasilyev F P 1980 *Numerical methods for solving extreme problems* (Moscow: Publishing House «Nauka») p 520

[4] Polyak B T 1983 *Introduction in optimiztion* (Moscow: Publishing House «Nauka») p 384

[5] Zhiglyavskii A A and Zilinskas A G 1978 *Methods of search for global extremum* (Moscow: Publishing House «Nauka») p 247

[6] Sergeev J D and Kvasov D E 2008 *Difgonal global optimiztion methods* (Moscow: Publishing House «Fizmalit») p 352

[7] Panteleev A V 2009 *Metaheuristic algorithms of search of global extremum* (Moscow: Publishing House «Aviation Institute-Print») p 159

[8] Leyffer S 2001 Integration SQP abd branch-and-bound for mixed integer nonlinear programming *J. Comput. Optim.* **18** 295–309

[9] Fletcher R and Leyffer S 1994 Solving mixed integer programs by outer approximation *J. Math. Program* **66** 327–349

[10] Geoffrion A M 1972 A generalized Benders decomposition *J. of Optimization Theory and Applications* **10(4)** 237–260

[11] Westerlund T and Pettersson T 1995 An extended cutting plane method for solving convex MINLP problems *J. Computers Chem. Eng.* **19** 131–136

[12] Mawengkang H and Murtagh B A 1986 Solving nonlinear integer programs with large-scale optimization software J. *Annals of Operations Research* **5** 425–437

[13] Rouban A I 2003 Method for global optimization in continuous space *J. AMSE Series: Advances A* **40: 4** 9–28

[14] Rouban A I 2004 *Global optimization by a method of averaging of coordinates* (Krasnoyarsk: State Technical University Publisher) p 303

[15] Rouban A I 2013 Global optimization method based on the selective averaging coordinate with restrictions *J. Control and Computer Science* **1(22)** 114-123

Fine structure of inelastic electron scattering cross-section spectra for MN

A S Parshin[1], A Yu Igumenov[1], Yu L Mikhlin[2], O P Pchelyakov[3,1], V S Zhigalov[4,1]

[1]Siberian State Aerospace University named after Academician M.F. Reshetnev
31 “Krasnoyarskiy Rabochiy” prospect, Krasnoyarsk, 660037, Russia.
[2]Institute of Chemistry and Chemical Technologies of the Siberian Branch of the Russian Academy of Sciences, 50/24 “Akademgorodok”, Krasnoyarsk, 660036, Russia.
[3]Rzhanov Institute of Semiconductor Physics of the Siberian Branch of the Russian Academy of Sciences, 13 “Academician Lavrentjev Av.”, Novosibirsk, 630090, Russia.
[4]Kirensky Institute of Physics of the Siberian Branch of the Russian Academy of Sciences,
50/38 “Akademgorodok”, Krasnoyarsk, 660036, Russia.

E-mail: aparshin2010@mail.ru

Abstract. The comparative analysis of the reflection electron energy loss spectra and the inelastic electron scattering cross-section spectra for Mn was carried out. It is shown that inelastic electron scattering cross-section spectra have certain advantages in the study of the interaction of electrons with the substance as compared to the electron energy loss spectra. The inelastic electron scattering cross section spectra fine structure was analysed by fitting the experimental spectra using the 3 parameters Lorentzian-type formula of Tougaard. This method was used for the quantitative analysis of the contributions of various loss processes in the inelastic electron scattering cross section spectra, determination of the loss peaks energies and origin.

1. Introduction

Mn is widely used in multilayer metal/semiconductor structures, which open opportunities for the development of micro-, nanoelectronics and spintronics [1, 2]. The study of multilayer structures physics, metal/semiconductor interfaces and silicidation processes are necessary for the development of nanotechnology. One of the most informative methods of research materials is electron spectroscopy, in particular reflection electron energy loss spectroscopy (REELS). From REEL spectra the inelastic electron scattering cross section spectra (so-called $K\lambda$-spectra [3]) can be calculated and allow to investigate the electron-solid inelastic interaction more detailed [2, 4-10]. Analysis and interpretation of the electronic spectra are important for understanding the physical processes of the investigating systems. Both the REELS and $K\lambda$-spectra for Mn are underinvestigated.

Inelastic electron scattering cross section spectra indicate the probability of electron energy loss T in a single scattering event. $K\lambda$-spectra can be used for quantitative determination of element concentrations in two-component composite structures. The ability of quantitative determination of atomic element concentrations in Fe_xSi_{1-x} [5, 6], Mn_xSi_{1-x} [2] and Ge_xSi_{1-x} [7] systems were investigated earlier with the dependence of the maximum value of $K\lambda$-spectra for the standard samples. Computer modeling of inelastic electron scattering cross section spectra in terms of the dielectric response theory was used for the investigation of elements depth distribution and for determination of silicon dioxide film thickness on a silicon substrate [6].

In this paper, the method of fine structure analysis of inelastic electron scattering cross-section spectra suggested by us in [9, 10] is applied to analysis of Mn $K\lambda$-spectra. This method is based on fitting inelastic electron scattering cross-section spectra to the 3 parameters Lorentzian-type formula of Tougaard [11] and used for determination the origin of the loss peaks and unresolved peaks analysis.

2. Experimental

The spectra for Mn were measured employing two multilayer systems composed of (1) Si substrate, 50 nm Mn film, 30 Å Pd layer protecting the Mn layer from oxidation and (2) glass substrate, 50 nm Mn film, and protecting 30 Å Ge layer. The samples were prepared using the thermal evaporation method under high vacuum conditions.

The experiments were carried out using a photoelectron spectrometer SPECS (Germany) equipped with the spherical energy analyzer PHOIBOS MCD9, double anode X-ray tube as an X-ray source, and a Microfocus EK-12-M electron gun (STAIB Instruments) for excitation of the REEL spectra. The surface contaminations, protecting layers and oxide layers were removed applying Ar^+ ion etching (accelerating voltage of 2.5 kV, ion current of 15 μA) with an ion source IQE-12/38 (SPECS) in the analytical chamber of the spectrometer before the REELS measurements; the completeness of the procedure was controlled via the relevant photoelectron and Auger spectra.

REEL spectra were acquired in the interval of 150 eV below the elastic peak with an energy step of 0.1 eV. Energy loss T was calculated as the difference between the primary electron beam energy E_0 (zero loss) and the reflected electron energy E, $T = E_0 - E$. The primary electron beam energies were 300, 600, 1200, 1900, 3000 eV, and the full width at half-maximum (FWHM) of the primary electron beam was less than 1 eV.

The inelastic electron scattering cross section spectra which are the products of inelastic mean free path λ and inelastic scattering cross section $K(E_0, T)$ (where E_0 is the primary electron energy, $T = E_0 - E$ is the electron energy loss and E is the reflected electron energy) were calculated from the experimental reflection electron energy loss spectra with the software package QUASES^{TM} XS REELS (Quantitative Analysis of Surfaces by Electron Spectroscopy cross section determined by REELS) [12] based to the algorithm suggested in [13].

3. Results and discussion

3.1. The comparative analysis of the reflection electron energy loss spectra and the inelastic electron scattering cross-section spectra for Mn

On the REELS for Mn (Fig. 1, a) there are two peaks: the intensive peak at energy 21.8 eV and less intense one at energy 51.5 eV, corresponding respectively to excitations of volume plasmon and core level excitation M_{23} in Mn. In the low-energy region the feature with energy of about 5 eV is observed.

For the more accurate peak energy determination and reducing of the impact of inelastic electrons background, numerical differentiation of the experimental curves $N(T)$ (N is the number of electrons with energy loss T) was carried out (Fig. 1, b). In the differential spectra for Mn in addition to the bulk plasmon and M_{23} peak, the peak at energy 9 eV is observed. The average loss peaks energies determined from the differential REELS are 10.3 ± 1.0, 21.7 ± 1.4 and 49.4 ± 0.2 eV. The surface plasmon is not resolved in these spectra. Peak energies are close to the literature data for manganese [2, 14-16].

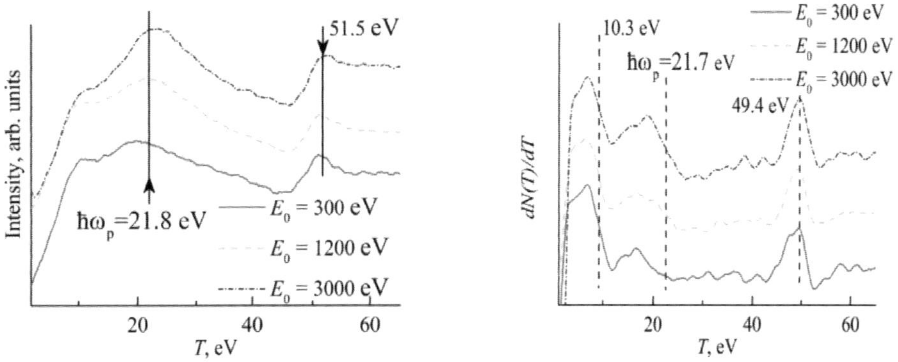

a) b)

Figure 1. Reflection electron energy loss spectra for Mn in integrated (a) and differential (b) species.

The inelastic electron scattering cross-section spectra for Mn obtained with primary electron energies 300, 1200, 3000 eV are shown in figure 2.

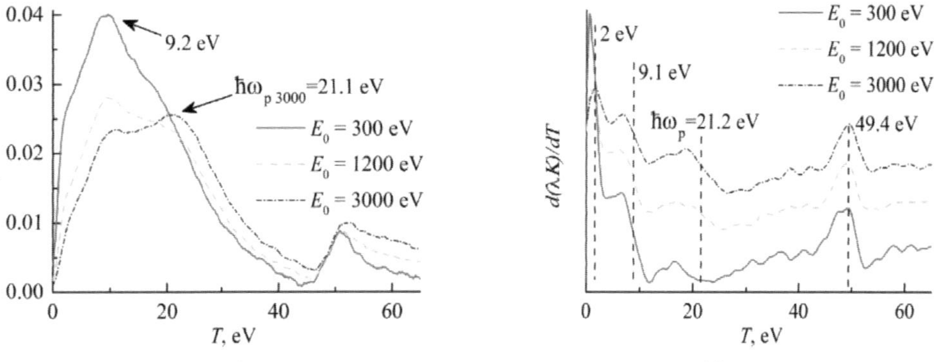

a) b)

Figure 2. Inelastic electron scattering cross-section spectra for Mn in integrated (a) and differential (b) species.

Inelastic electron scattering cross-section spectra in contrast to REELS have absolute units of energy loss. At figure 2 (a) the variation of spectrum shape with primary electron energy is observed: at E_0 = 300 eV the peak at 9.2 eV is dominating and it decreases with E_0. At E_0 = 3000 eV the peak at 21.1 eV, corresponding to volume plasmon loss in Mn [2, 14, 15], is dominating. It is worth noting, that the $K\lambda$-spectrum shape is more affected by the variation of primary electron energy that the REEL spectra. This is probably generated by the increasing of surface loss intensity in consequence of the subtraction of multiple electron scattering background according to algorithm [13]. At these spectra the M_{23} peak is always observed. At figure 2 (b) the differential $K\lambda$-spectra for Mn is shown. Differentiation allows us to determine the energy of low-intensity peak at 2 eV, but surface plasmon is still not resolved. The average loss peaks energies determined from the differential $K\lambda$-spectra are 2.0 ± 0.5, 9.1 ± 0.2, 21.2 ± 1.8 and 49.4 ± 0.2 eV. These energies are closed to loss peaks energies obtained from REELS.

3.2. The fine structure of inelastic electron scattering cross-section spectra analysis

The variation of inelastic electron scattering cross section spectrum shape with the primary electron energy is affected by the variation of surface-like and volume-like excitation intensities. In Ref. [9, 10] the method of inelastic electron scattering cross-section spectra fine structure analysis was suggested. The fine structure analysis is based on fitting $K\lambda$-spectra to the 3 parameters Lorentzian-type formula of Tougaard [11]:

$$\lambda K = \frac{BT}{\left(C - T^2\right)^2 + DT^2}.$$

B, C, D are fitting parameters and have special values for different elements [11]. The peak intensity depends on parameter *B*, peak energy depends on parameter *C*, peak width and, indirectly, peak energy depend on parameter *D*.

We chose the 3 parameters Lorentzian-type formula of Tougaard because it describes the inelastic electron scattering mechanisms. The $K\lambda$-spectra fitting to the Tougaard peaks can be used to analyze different mechanisms of inelastic electron scattering such as volume plasmon excitations, surface plasmon excitations and interband transitions.

The results of fitting the Mn $K\lambda$-spectra that were obtained with primary electron energies 300 and 1900 eV is shown in figure 3.

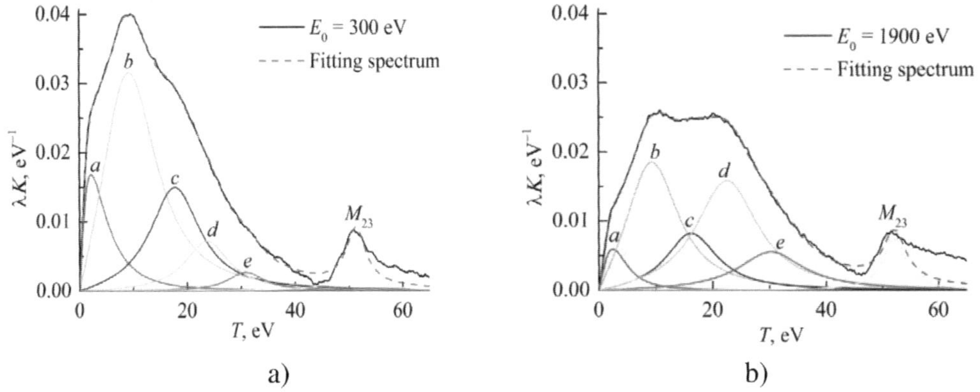

Figure 3. The inelastic electron scattering cross-section spectra for Mn obtained with primary electron energies 300 (a) and 1900 (b) eV fitted to the 3 parameters Lorentzian-type formula of Tougaard.

The number of fitting peaks was chosen to provide the minimal least-square deviation of the fitted spectrum from the experimental spectrum. Six peaks including peak M_{23} are sufficient to provide a good coincidence between the experimental and fitting spectra. The inelastic electron scattering cross-section spectra for Mn were approximated with 6 Tougaard peaks *a, b, c, d, e* and peak M_{23} (core level excitation [2, 14, 16]). Parameters *B, C, D* of the 3 parameters Lorentzian-type formula of Tougaard were varied for the best least squared fit to the experimental spectra.

The primary energy dependences of the fitted peak areas (that describes the excitation intensity) are shown in figure 4. The gradual decrease with primary energy areas of peaks *b* and *c* indicate their surface origin (surface-like peaks); the opposite primary energy dependence of the peak *d* area indicates its bulk origin (volume-like peak). The other peaks origin is hard to determine because of their small intensity.

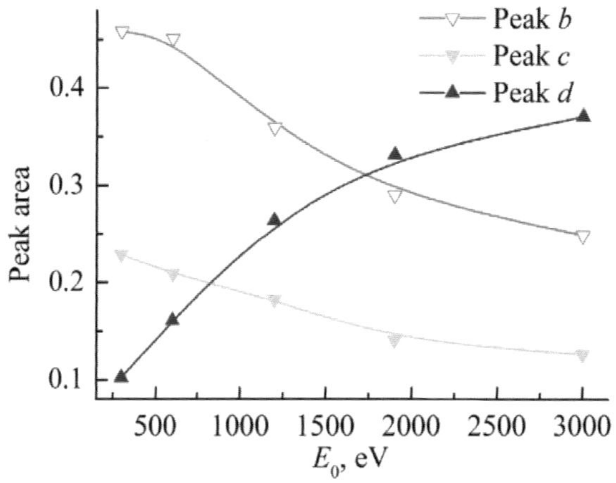

Figure 4. The primary energy dependences of the fitting peaks area

The average energies of the fitted peaks and their root-mean-square deviation are shown in Table 1. The fine structure analysis method allows determining the peaks energies balancing out an effect of summation of different losses intensities to the energies of spectrum maximums.

Table 1. Fitting peaks energies

Fitting peak	*a*	*b*	*c*	*d*	*e*	M_{23}
Fitting peak energy, eV	2.5 ± 0.4	9.2 ± 0.1	16.9 ± 0.8	23.0 ± 0.6	30.6 ± 0.6	51.8 ± 0.4

The volume-like peak *d* corresponds to volume plasmon because it has gradually increasing primary energy dependence of area and its energy loss is close to the bulk plasmon energy in Ref. [2, 14, 15] (Table 1). Peak *c*, corresponds to a surface plasmon because it has gradually decreasing primary energy dependences of area and the energy loss ratio between the volume plasmon and the peak *c* is 1.35 that is quite close to the well-known $\sqrt{2}$ from the classical plasmon model [15, 17]. This peak was unresolved in REELS and $K\lambda$-spectra for Mn but the method of fine structure analysis allows determining its energy, intensity and origin. The M_{23} peak related to the electron transition from the level M_{23} and its energy loss is close to the literary data for Mn [2, 14, 16].

This method is look like the method of fitting the XPS spectra to the peaks. The functions used in XPS-spectra fitting are Gaussian-Lorentzian curves (Voigt-functions) and Doiach-Sunjic functions [18]. The peak fitting in spite of some difficulties with mathematical uncertainty became an important tool in XPS [18, 19] that is used for the chemical binding determination. The peak fitting in inelastic electron scattering cross section spectroscopy can be used for the investigation of electron scattering mechanisms and quantitative determination of the different processes contributions to the inelastic electron scattering cross section spectra.

4. Conclusions

The reflection electron energy loss spectra and the inelastic electron scattering cross-section spectra for Mn were investigated. Comparative analysis of these spectra shown, that the inelastic electron scattering cross-section spectra are more sensitive to changes in the primary electron energy and allow to analyze, not only energy, but also the intensity of the losses. To investigate the fine structure for Mn inelastic electron scattering cross-section spectra the method based on its approximation with 3 parameters Lorentzian-type formula of Tougaard was applied. Using the fine structure analysis method the investigation of the unresolved loss peaks, determination of its energies and origin were carried out.

References

[1] Ohtsu N., Oku M., Nomura A., Sugawara T., Shishido T., Wagatsuma K. 2008 *Appl. Surf. Sci.* **254** pp 3288-3294.

[2] Parshin A. S., Pchelyakov O. P., Dolbak A. E., Ol'shanetskii B. Z., 2013 *Journal of Surface Investigation. X-ray, Synchrotron and Neutron Techniques* **7** pp 505-508.

[3] Gergely G. *2002 Prog. Surf. Sci.* **71** pp 31-88.

[4] Jin H., Shinotsuka H., Yoshikawa H., Iwai H., Tanuma S., Tougaard S. 2010 *J. Appl. Phys.* **107** pp 0837091-08370911.

[5] Parshin A. S., Aleksandrova G. A., Dolbak A. E., Pchelyakov O. P., Ol'shanetskii B. Z. , Ovchinnikov S. G. , Kushchenkov S. A. 2008 *Tech. Phys. Letters* **34** pp 381-383.

[6] Parshin A. S., Kushchenkov S. A., Aleksandrova G. A., Ovchinnikov S. G. 2011 *Tech. Phys.* **56** pp 656-661.

[7] Parshin A. S., P'yanovskaya E. P., Pchelyakov O. P., Mikhlin Yu. L. , Nikiforov A. I., Timofeev V. A. , Esin M. Yu. 2014 *Semiconductors* **48** pp 224-227.

[8] Parshin A. S., Kushchenkov S. A., Pchelyakov O. P., Mikhlin Yu. L., Hasanov T. 2012 *Optoelectronics, Instrumentation and Data Processing* **48** pp 396-400.

[9] Parshin A. S., Igumenov A. Yu., Mikhlin Yu. L., Pchelyakov O. P., Nikiforov A. I., Timofeev V. A., 2015 *Semiconductors* **49** pp 423-427.

[10] Igumenov A. Yu., Parshin A. S., Mikhlin Yu. L., Pchelyakov O. P., Nikiforov A. I., Timofeev V. A. 2014 *Vestnik SibGAU* **4** pp 230-235.

[11] Tougaard S. 1997 *Surf. Interface Anal.* **25** pp 137-155.

[12] Tougaard S. QUASES – Software packages to characterize surface nano-structures by analysis of electron spectra http, available at: http://www.quases.com/.

[13] Tougaard S., Chorkendorff I. 1987 *Phys Rev. B* **35** pp 6570-6577.

[14] Wehenkel C., Gauthe B. 1974 *Phys. stat. sol. (b)* **64** pp 515-525.

[15] Lifshits V. G., Lunyakov Yu. V. 2004 Electron Energy Loss Spectra of Surface Phases on Silicon *Vladivostok: Dal'nauka.*

[16] Feldman L. C., Mayer J. W., 1986 Fundamentals of Surface and Thin Film Analysis *New York: Elsevier Science.*

[17] Raether H. 1988 Surface Plasmons on Smooth and Rough Surfaces and on Gratings *Heidelberg: Springer-Verlag Berlin* 135 p.

[18] Gomoyunova M. V., Pronin I. I. 2004 *Tech. Phys.* **49** pp 1249-1272.

[19] Peak Fitting in XPS, available at: http://www.casaxps.com/.

On the question of economic efficiency and how to assess it

M O Petrosyan[1], I V Kovalev[1], P V Zelenkov[1], M N Chuvashova[1], I A Grishina [2], K K Pershakova[1]

[1]Siberian State Aerospace University named after Academician MF Reshetnev
31 "KrasnoyarskiyRabochiy" prospect, Krasnoyarsk, 660037, Russia.
[2]Russian University of Economics named after GV Plekhanov. Krasnoyarsk branch

E-mail: mopetrosyan@gmail.com

Abstract. Over the years, since the world moved to a system of market-based economic relations scientists and businessmen worldwide wonder how to ease and modernize the process of analyzing business performance. During this time, a variety of methods and tools have been developed to deal with this problem. After all, the productivity of the company is directly related to it's competitive qualities, so the managers are interested in the quality of evaluation of their companies. The article deals with the traditional methods of evaluating the performance of enterprises and some of the tools used in order to improve its quality.

The economic efficiency is complex category of economics. The evidence of it's presence in all spheres of life and all stages of economic production are incontestable. It is the basis of quantitative criteria for constructing value of decisions. In general, the representation of the efficiency characterizes the development of systems, processes and events. Efficiency advocates both as an indicator of the development of the enterprise and the most important stimulus of s growth and development. Enterprises realize their mutual interests and operating in a competitive environment, occupying a certain segment of the market. The increase in market share, the increase of competitive advantage cause stability of strategic positions of organization, that is characterized by the ability to function effectively in the long term.

Efficiency broadly describes the qualitative aspect of development of enterprise by showing what combination of public resources helped to receive final result. Quantitatively, the efficiency is expressed by the ratio between the results obtained in the production process, and the costs of social labor, associated with the achievement of these results. Therefore, the meaning of increasing of efficiency production is a more rapid growth of the result (effect) as compared to the cost.

An analysis of the overall effectiveness of economic activity is the prerogative of senior management structures of the organization.

Analysis of the effectiveness of the company needs not only its head to assess the financial situation, but also a number of people who are directly involved in economic practice:

- Investors, who make decisions about formation of a portfolio of securities companies;
- Creditors, who if necessary, should give loans to businesses;
- Auditors, who check the statements and the economic activity of enterprises;
- Heads of marketing and advertising departments, who based on this information, create strategy of moving goods to markets.

Modern economy, with its ever-increasing competition, requires continuous improvement of the efficiency of the economic potential from managers of enterprises, firms and associations. The best forms of management are also important in order to achieve the best results at the lowest cost of labor and resources. The need to improve the theory and practice of the analysis of the economic efficiency of enterprises further determines the relevance of the article. As a characteristic of the quality of its operations and, consequently, its viability (in the market environment is manifested through competitiveness), efficiency is not only the most important area of theoretical and empirical research in economics and management, but also the object of close attention as for the management of the organizations as well for other interested parties.

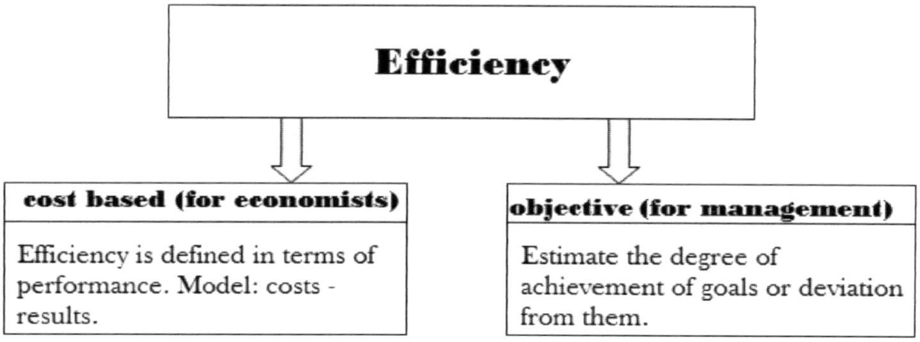

Figure 1. Types of efficiency

Currently, the most complete and consistent economic research of efficiency of economic activities is given in the theory of complex economic analysis where large areas are dedicated to long-term effectiveness, operational and current analysis, which is estimated to reveal changes, unrealized potential and increase effectiveness of economical activities of the enterprise on the basis of changes in efficiency of economic activities.

An analysis of the overall effectiveness of economic activity is the prerogative of the senior management and linked to the definition of product price, lot size of purchases of raw materials and supplies products, equipment replacement or technologies. Other solutions must also be assessed in terms of overall success of company, the nature of its economic growth and efficiency. Main objectives of efficiency analysis are the following: assessment of the economic situation; detection factors and causes of the state of progress; preparation and justification to management decisions; identification and mobilization of reserves.

In conditions of market relations in the economy, business analysis is designed to provide high profitability and competitiveness of organizations both in the short and in the long term.

Under the new economic circumstances, deeper understanding the content of the concept of efficiency is necessary. The level of efficiency of production does not reflect only the ratio of benefits and costs, but also the extent to which economic potential is used. The overall aim of producers, of course, is income. Therefore, as a local criteria of efficiency is considered to be the ratio of profit to the cost of production, that is the traditional indicator of profitability. However, from the standpoint of system-wide objectives, role assignment is producing the set (the range) of products of a certain quality and amount and in a predetermined time interval.

Therefore, from a system perspective, the effectiveness of any producers (enterprise, organization) should be measured by its contribution to the system benefits.

The method of economic analysis includes a whole system of methods and techniques, which allows to scientifically study economic phenomena and processes that make up the economic activities of the organization. At the same time any of the methods and techniques used in the economic analysis can be called "method" in the narrow sense of the word as a synonym for "the way to do". Analysis of economic activity also uses the methods and techniques that are typical for other sciences, especially mathematics and statistics.

The method of analysis is a set of methods and techniques, needed to ensure systematic and comprehensive study of the impact of individual factors on the change in the economic performance and ways to improve the activities of organizations. There are lots of different practices, used in the analysis of economic activities. These include the traditional methods, which are widely used in other disciplines for the processing and review of information (comparison, graphic, balance, average and relative numbers, analytical groups).

Comparison is the most common method, which allows the manager or evaluator to see the dynamics of enterprise development and make appropriate conclusions. Even if other methods of assessment were taken as a basis , the final stage always uses comparisons. The results are compared with the results of any other periods, or with the results of other similar companies operating in the market. Also, the results can be compared with the general trends in the business environment.

Analysis of socio-economic processes and the factors, which can influence them, often requires the investigator to comprise and rank these processes and factors. One of the main problems encountered with this type of analysis is that the criteria by which it is held,are often heterogeneous. This may be, for example, both financial and non-financial performance and the results of sociological research, and expert assessment.

In economic and management studies an object of evaluation is usually assumed to be the separate processes (often industrial), operations (such as promotions), projects (such as innovation) implemented by organizations or their structural units (divisions). Thus these objects (organization units) describe a certain set of characteristics that are interpreted as parameters or variables of its activities.

With the operationalization of the concept of efficiency of the organization, it is important to distinguish between performance (effectiveness) and efficiency of its activities. The first one is the measure of the degree of achievement of the objectives set for the organization. The second, by comparing the obtained results with the analyzed unit values of factors (usually resource costs) under which they were achieved, gives an idea of how efficiently it operates. The most common understanding of the effectiveness includes both performance and efficiency (productivity) of the organization. It corresponds to the notion of efficiency when it comes to the correlation of the result to ensure its cost to the organization.

Measurment the effectiveness of the organization through the analysis of the productivity of its operations for the respective technological display "Input - Output", leading to the presence in the resulting, estimates a number of universal properties, which are not associated with any particular type of organization, or with the specific activities in relation to which valuation. Such properties include:

a) the system (synthesis) nature or integrity of this estimation;

b) qualitative and quantitative certainty of a numerical or ordinal scale;

c) the specific content, which is specified in accordance to the nature of the activity in considered organizational unit.

Nowadays, a variety of tools exists and still develops in order to help in the evaluation of the company. In a relatively short period of time Data Envelopment Analysis (DEA) has grown into a powerful quantitative, analytical tool for measuring and evaluating performance of organization. It is a methodology of comparative analysis of complex technical, economic and social systems.

DEA it is a method based on linear programming, needed to determine the relative effectiveness of different means implemented enterprise products. Data Envelopment Analysis is a relatively new "data oriented" approach for evaluating the performance of companies in various aspects. Linear programming is the underlying methodology that makes DBA particularly powerful, comparing with alternative productivity management tools. DBA has been widely studied, used and analyzed by academics that understand linear programming. [Springer]

That may be economy or management of organization or some other aspects. DEA also can be used to make adjustments and additions to the assessment carried out earlier by other methods and means. Data Envelopment Analysis has been recognized as a valuable analytical research instrument and a practical decision support tool.

In fact, the DEA method is a tool for benchmarking, the use of which not only allows you to set the most effective organizational units and select one of them according to the standards of inefficient facilities, while identifying how far the latter are located from the first.

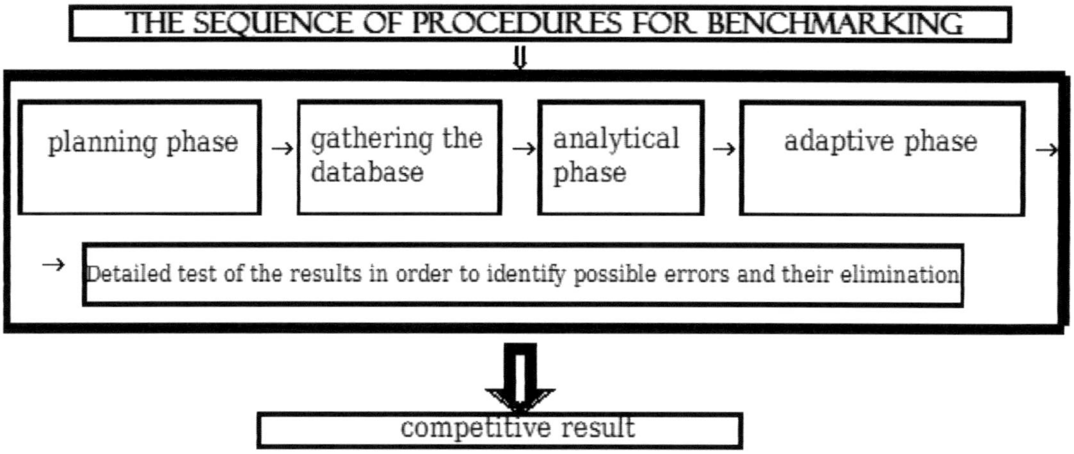

Figure 2. Benchmarking scheme

As shown in the scheme, benchmarking aims to improve the competitive qualities of companies, in a same way that pursues the head of the organization conducting the analysis of its effectiveness.

A key aspect of DEA is incorporating environmental factors into the model as either inputs or outputs. Resources available to units are classified as inputs whilst activity levels or performance measures are represented by outputs.

According to the formula, the efficiency is given by:

$$\text{Efficiency} = \text{output}/\text{input}$$

Thus, we get the ratio by which we can judge the effectiveness of a working unit. If each working unit achieved the highest profits in relation to the costs, the absolute or optimum efficiency will be reached. This means that it is impossible to achieve the best performance results without introduction of new technology or other variations in the manufacturing process.

Another tool of practical interest to assist the analysis of the effectiveness of the enterprise is the method TOPSIS (Technique for Order Preference by Similarity to Ideal Solution). TOPSIS is based on the principle, that the object, which showed the best results, is being as close as possible to the best object and as far from the worst of the objects. The best / worst object - it is hypothetical object for which the value of all the analyzed indicators are the best / worst. Thus, TOPSIS allows to determine the position of the objects on the best and worst scale, as well as relatively to each other. TOPSIS assumes that we have "m" alternatives (options) and "n" attributes/criteria and we have the score of each option with respect to each criterion. TOPSIS method is an effective tool to assist decision-makers and experts in formulation of their goals and subjective preferences, structuring of the set criteria, evaluation of the alternatives in the decision-making process in the language of fuzzy mathematics, linguistic variables, fuzzy sets and fuzzy numbers.

Speaking on the analysis of the effectiveness of the enterprise, the decision-makers should always keep in mind that no matter how many good tools exist to help you to carry it out, the final decision will still always be taken by manager. Whatever method he decided to take, or tool he decided to choose to help, the result of the analysis will depend not on the helping practices, but on the man himself. And the degree of professionalism will be a key factor in a process of summing up the results of the analysis and its interpretation for further development of enterprise development strategies.

References

[1] Mamedova M G Dzhabrailova Z G. 2015. Multi-criteria optimization of human resources management based on a modified method TOPSIS. *East Eur. J. of adv. Tech.* №74 p. *48-62*

[2] Fedotov Ju V. 2012. Method DEA: data analysis convolution. *Russian Management Journal.* №2 p. *48-62*

[3] Kelemenis, A. Askounis D. 2010. A new TOPSIS-based multi-criteria approach to personnel. *Exp. Sys. with Applications.* Vol. 37, Issue 6. P. 4999–5008.

[4] Fedotov Ju V. 2011. Analysis of existing methods of estimation of the enterprise effectiveness. *Bulletin of LSU. AS Pushkin.* Vol. 4, Issue 6. p. 16-24

[5] Yisau Abiodun Babalola. 2015. Analysis of existing methods of estimation of the enterprise effectiveness. *LAP*. P. 168

[6] http://deazone.com/en/resources/tutorial/issues-in-dea

Information support for high technologies: issues of innovation

S Yu Piskorskaya, A E Goncharov, G A Prohorovich, A V Perantseva
Siberian State Aerospace University
31 Krasnoyarskiy Rabochiy prospect, Krasnoyarsk, 660037, Russia

E-mail: piskorskayasyu@sibsau.ru

Abstract. The current development of high technologies and innovative projects requires systematic information support. This article describes examples of information support and promotion of regional technological platforms of the Krasnoiarskii krai on the base of communications projects which are being realized by students at SibSAU. These technological platforms correspond to the prioritized fields of developing science and research in the Russian Federation.

Keywords: high technologies, innovative projects, information support, Krasnoiarskii krai, SibSAU

Development in the economic and social field is determined today by the percentage of high-tech production facilities and the realization of innovative projects. This statement is especially true for Russia, which has been lagging behind the more economically developed countries. Such projects are realized based on scientific research, progress made in technological industries. The key result of their implementation is a high-quality novel high-tech product possessing unique properties; it can also be a new technological process that is extremely efficient in terms of production and costs [1, 2].

It is obvious that for such projects to be launched and realized it is necessary to have in the industry not only material resources, but unique personnel – researchers, scientists, high-profiled specialists, managing staff.

Due to the fact that the role of a country in global technological space is determined by its scientific and engineering achievements together with the implementation of this scientific knowledge in industry [3], it is necessary to provide decent information support for all innovative projects. Such support benefits social and economic relations between the government and business; it becomes possible to swiftly react to fluctuations on the market, forecast demand for projects and technologies, determine the addressed target audience, and arrange new contracts. It also enables to inform the general public of the efficiency of major innovative projects with minimal loss from their realization; this may include even such factors as the reduction of pollution due to the exclusion of paper handouts and other advertisement products.

In modern Russia the principal inclusion of the innovative paradigm into all state official documents on territorial development, accepting it as a key aspect for the modernization of the economy has resulting in the establishment of specialized centers. They are known as regional technological platforms (RTPs) and correspond to one of the prioritized fields of science and development in the Russian Federation.

An example is the strategy for the development of the Krasnoiarskii kai: *The innovative krai – 2020* [4]. In this strategy there are 13 priority activities for the regional technological platforms. There strategic target is the development of innovative technologies, which are focused on improving the attractiveness of residing in the Krasnoiarskii krai.

It should be noted, that unlike European or Russian federal platforms, the initiators of establishing technological platform in Krasnoyarsk are universities: Reshetnev Siberian State Aerospace University, Siberian Federal University, Krasnoyarsk State Medical University, Krasnoyarsk State Agricultural University, and Siberian State Technological University.

The reason for this is the presence of a considerable scientific and technical potential at universities, its readiness to offer new technologies and technical solutions to enterprises. Besides, universities are supplies of high-qualified professionals; these may act as a consolidating force for major and middle business companies since they are alumni of a particular educational institution.

It should also be considered that the academic community is the more advanced part of society when it comes to innovative transformations [5]. Thus, the university has become a platform for developing minor innovative enterprises (MIEs). Products created in these enterprises were they underwent approbation in case of success may be put into serial production. This requires correct information support for their advertisement on the market. Taking into consideration that the communications sector is subjected intense social, economic, and technological transformations, it should be noted that professionals working in the field of information support for technological projects should be aware of the significant increase of information flows. Another aspect is the growing competition for attention among the general public. The development of digital communications, including the whole sector of new media and social networks, along with the increasing demand for anti-crisis communication all play a key role in this phenomenon [6].

As a result, requirements for universities also experience a certain degree of transformation: the requirements become more flexible and sensitive to social and technological challenges, they are aware of competition and innovation. These demands require universities to consider a specific competence which is scientifically explained in their work. These competences also agree to the social and economic situation in the country and the region. In this situation communications technologies become an effective tool for the promotion of innovative activity.

At SibSAU a number of projects that are considered as serious technological research which have undergone international approbation is being realized. These projects are directly related to the promotion of innovations, technological projects, and perspective ideas [7, 8].

The development of such ideas can suggest new ideas for research, the solution of practical problems, related with regulating the public and individual opinion for the realization of sociocultural, scientific and educational, and innovational-economic programs of regional and national scale.

This direction in the development of national project corresponds to the priorities in state support for scientific, technological, and innovative activity in the Krasnoiarskii krai which has been ratified by the Number 8-3635P Act of the regional legislative assembly of the Krasnoiarskii krai from 7 July 2009 – *The economic, social and legal aspects of the development of the Krai*. It also agrees with the *List of prioritized fields of forming regional technological platforms* signed by the governor of the Krasnoiarskii krai on 24 November 2011, assigned number 218-ug.

Besides these official documents, the aforementioned program is proposed according to the *Strategy of innovative development of the Krasnoiarskii krai to the year 2020*; this strategy influences the main aspects of increasing the potential of human resources in science and research, education, technologies. Its scope is to provide sophisticated conditions for the development of scientific schools adjacent to universities, provide support for organizing a high level of education in engineering, management, and humanities; this education should be directly related to the development of innovative systems in the Krasnoiarskii krai.

The development of technological platforms requires systematic propaganda in order to form a positive public opinion towards hi-tech industries and hazardous production facilities. Hence, among the main aims of student research projects in the field of information support for hi-tech innovational projects, it would be obvious to see a project which would study the forming of a positive public opinion towards critical technologies and industrial facilities which are being realized in the Krasnoiarskii krai. In order to achieve the proposed goals it is necessary to solve the following problems:

1. Overcoming any existing negative stereotypes and myths about the danger of critical technological and engineering facilities.

2. Popularizing the achievements of national nuclear and space research, innovative technological projects and high technologies utilized in these fields.

3. Popularizing the achievements of regional industries and engineering facilities in the field of high technologies.

These problems are united under the idea of popularizing scientific knowledge. Its main goal is to inform the target audience about complex innovative ideas in simple ways, including such methods as audiovisual creative projects.

As possible expected effects from introducing the results of realizing these projects it would be logical to focus the economic (financial support), social (decreasing social tension), and educational (forming a specific "polygon" for testing new communications technologies) effects. Among the more significant factors it is necessary to mention the forming and harnessing a favorable public opinion for a positive vision of the benefits of nuclear energy. A look at Russian history shows that public opinion towards this issue underwent diametrically opposite transformation, particularly after the Chernobyl disaster of 1986. Before Chernobyl the majority of the Soviet population saw nuclear energy as a safe and progressive source of energy. The Perestroika and Glasnost' era revealed other nuclear disasters that had happened in the USSR prior to 1986. This changed the public opinion towards nuclear energy. This transformation is still visible today. So, creating a positive vision of nuclear energy is essential to the nuclear energy industry, at the same time such propaganda increases the responsibility, in terms of environmental protection, of the industry for such activity.

Thus, communications technologies can be an effective tool for promoting hi-tech and innovative projects and providing them with information support. At the same time, they allow university students to approbate educational, information, ethical, and sociocultural technologies.

References

[1] Kizim A A 2013 Innovations as a key vector of logistical systems development *Teoriia I praktika obshestvennogo razvitiia*. No. 1

[2] Shishkina N A 2012 The category of "quality" for innovative projects *Vestnik of Siberian State Aerospace University* No. 6(46) P 240

[3] Sirotkin O S 2002 The role of the state in forming competitive hi-tech corporations *Investitsionnye proekty i tekhnologii* (Moscow Russia) p 29-32

[4] Strategy of innovative development of the Krasnoiarskii krai to the year 2020 "Innovative krai – 2020" http://www.krskstate.ru/innovation/policy/strategy.

[5] Tolstoi D A, Khramenko S A, Zakhar'in K N, Krasnodubova O N 2012 The aims of regional technological platforms of the Krasnoiarskii krai *Vestnik of Siberian State Aerospace University* No. 6(46) p 2015-216

[6] The situation in the national public relations industry. Report of the Russian academy of public relations http://www.slideshare.net/omukovozov/raos-pr-2014 p 8-9
[7] Mikhailov A V Slizhenkova D F, Mikheeva A R, Bybina Ia Iu 2012 Support and promotion of innovations in the system public relations tools *Vestnik of Siberian State Aerospace University* No. 6(46) p141
[8] Piskorskaia S Iu 2014 Tendencies in modern research in the field of communications at technical universities *Proceedings of the 18th international conference Reshetnevskie chteniia* Vol 3 p 233-235

Plasma chemical silicon etching process*

K V Rudenko[1], A V Miakonkih[1], A E Rogojin[1], S V Bogdanov[2], V G Sidorov[2], I V Kovalev[2]
[1] Institute of Physics and Technology of the Russian Academy of Sciences, Moscow, Russia
[2] Reshetnev Siberian State Aerospace University, Krasnoyarsk, Russia

E-mail: kovalev.fsu@mail.ru

Abstract. The cryogenic plasma chemical silicon etching process is developed, able to form the structured silicon layer with system of deep holes with high aspect ratio.

Keywords: betavoltaic element, p-i-n diode, cryo process, scalloping.

Introduction

The articles [2-5] studies silicon diode structure characteristics and their production methods, applicable for neutron detector creation. Shown, that planar epitaxial diode structures with narrow junction field region has leakage at the level ~ 1 uA/cm^2. In the same p-i-n diode structures with wide junction field region, leakage can be obtained at the level of 20 nA/cm$^{2.}$

It would be logical to assume, that if the structured silicon layer with formed on surface contiguous p+ layer gets to depletion area, it can not increase the leakages. It is clear that an area of junction will be greatly grow. Such variant of the structure was simulated in [1], that it confirms of idea about possibilities of the location of all structured silicon layer in the field of p-n junction, and maintaining leakage current at low level, that is typical for the planar p-i-n structure.

In this research, authors decide one of the basic task: research about possibilities of the high-aspect etching for forming of the structures with the developed geometry.

It was investigated some possibilities of the deep plasma chemical silicon etching to forming p-n junction on a surface of walls of the created structures.

Methods of creation silicon structures with high aspect ratio

Usually for silicon structures with deep holes used Bosch-process of plasma-chemical etching, widespread for production of microelectromechanical and nanoelectromechanical systems (MEMS and NEMS).

It is based on two-step cyclic process of plasma etching with a step of forming structure walls passivation and a step of its anisotropic etching. As a result, holes could be formed with a depth up to hundreds of microns in silicon. However, this method has some features, that can not always be combined with the requirements of micro- and nanoelectronics.

In particular, it is inevitable effect of scalloping walls, i.e. a type of roughness, which has a form of regular circular grooves (Figure 1).

* This research has been financed by the Ministry of Education and Science of the Russian Federation (contract No. 14.577.21.0117, unique project identifier RFMEFI57714X0117).

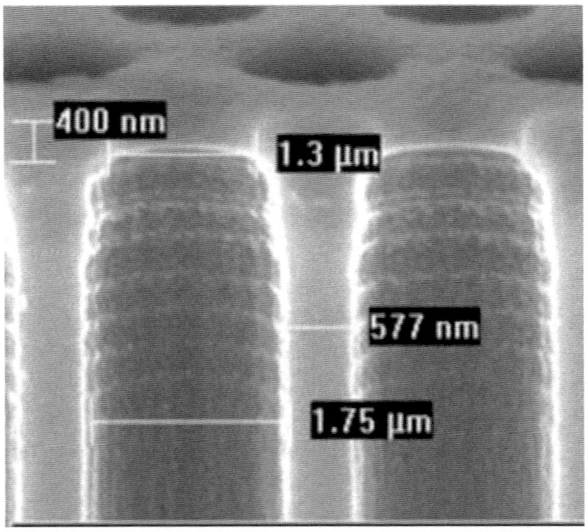

Figure1. Typically cross-section view of etched, Bosch-process.

This disadvantage could be flattened by reducing the duration of steps; however, it is impossible to completely get rid of it. If one form p-n junction on such surface side, it would be inevitably lead to increase the leakages in sharp places because of the field-effects.

The second feature is stubborn remnants of a thin polymer film in the holes (~1-3 nm). It polluting surface and has the potential opportunity to affect on the electrical characteristics of this structure. Therefore, in this research, we investigated the possibilities of another process, in particular the cryogenic process of deep plasma-chemical etching of silicon. Using of the cryogenic etching permits to get the structures with high aspect ratio and smooth walls. This method permit to use a plasma SF_6O_2 without gases that produce a polymers (such as different freons) and without modulation plasma composition during the etching process.

In particular, low temperatures (from -100°C to –130°C) permit to eliminate the isotropic spontaneous etching of the structure sidewalls by atomic fluorine, which carry out with purely chemical isotropic mechanism. As the result, it has the Arrhenius temperature dependence. Besides, for the same reason that one can achieve high silicon etching selectivity relative to the rigid mask of photoresist or silicon dioxide. During etching, oxygen formes protective coating composition SF_xO_y on the sidewalls, which prevents spontaneous etching by fluorine. The continuous process of cryogenic silicon etching is completely devoid of effect "scalloping", that permits to get etching structure with a minimum roughness of the walls. In this paper, wall roughness after etching of silicon in cryoprocess was investigated by authors.

An additional advantage of this process is the potential opportunity to permit managing etching profile (angle of inclination a wall). For etching on the specified depth is necessary apply highly selective lithographic mask. Use of an aluminium oxide mask was propossed. In the fluorine-containing plasma, the aluminium oxide forms non-volatile aluminum fluoride, which ensures high selectivity etching process.

Mask of aluminium oxide was deposited by atomic layer deposition method. As the organometallic precursor was used trimethylaluminium; as the oxidative precursor was used oxygen plasma.

To product the structures was used templates for contact optical lithography with «honey comb» structure with 5 um holes in diameter. Deep etching performed with the chamber with cryogenic table through the mask of aluminium oxide with autopsied windows. The plate temperature was maintained at -110°C.

IOP Conf. Series: Materials Science and Engineering **122** (2016) 012028 doi:10.1088/1757-899X/122/1/012028

Firstly, the technological process was fine-tuned on etching holes, as it is easier to get a cross-section image. The etching result shows in a figure 2. The image shows that the obtained vertical walls have depth of 40 um, with a small positive angle of 89 deg., the size's offset does not exceed of 20 nm. The selectivity with a respect to the mask is better than 1: 30000. The wall roughness study shows that this quantity is at most of the surface of the walls, with the exception of adjacent to a mask fields, and doesn`t exceeds the resolving power of scanning electron microscope (SEM), i.e. it is in the range 1-3 nm.

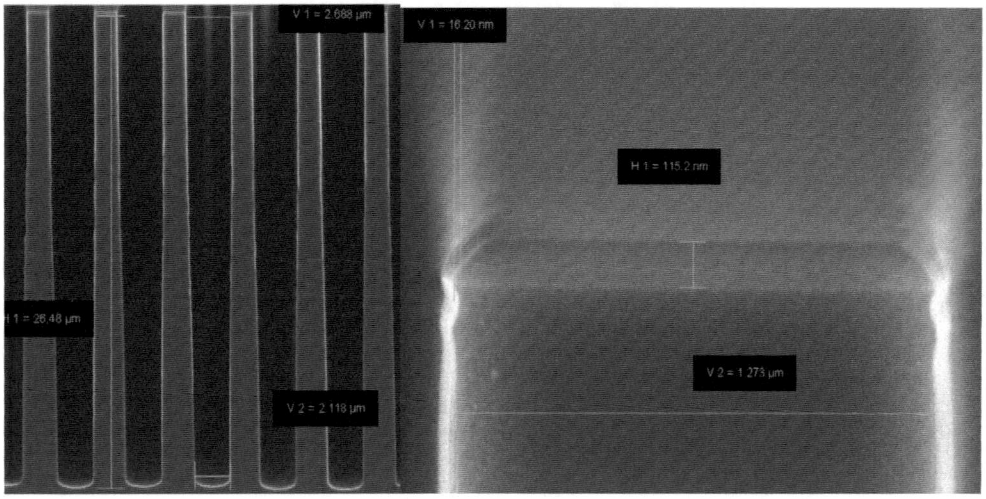

Figure 2. Cross-section view of etched by cryogenic process holes.

Designed cryo process was used for etching the structure with holes (Figure 3). The image shows that the desired structure holes without distortion geometric dimensions was achieved. Unfortunately, for this pattern is difficult to get a split, which perpendicular to the surface, and observe the etching profile without geometric distortion. However, carried out investigations permit to estimate the depth of the etching, offset of the sizes and the verticality of hole's walls.

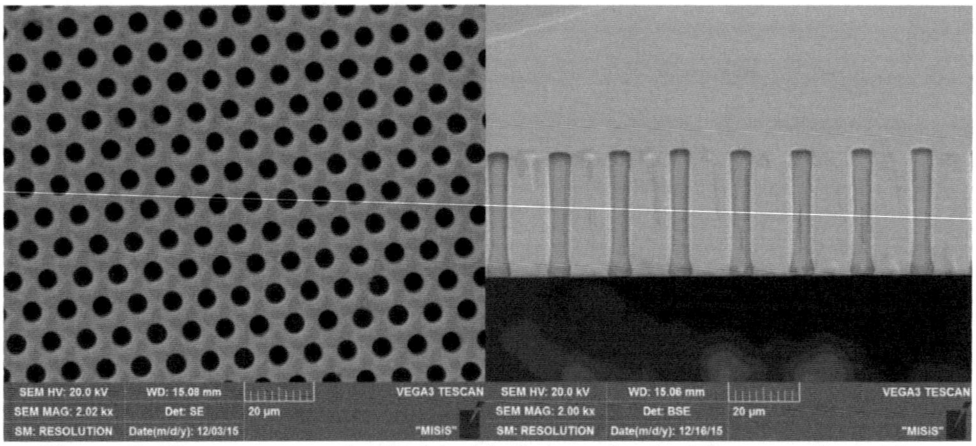

Figure 3. The structures of cylindrical holes, received with cryogenic etching (View up- left, split – on the right).

Further progress in the betavoltaic power source creation consist in practical realization of the formation of the radioactive ^{63}Ni continuous layer on the inner surface of the holes. The process technology of not radioactive nickel deposition into the pores is known and described in detail [6].

Conclusion

It was developed a cryogenic process of plasma-chemical etching silicon, which permits to develop the structured silicon layer, with a system of deep holes (5 um in diameter) with a depth of 45 um. The silicon layer has the necessary parameters for creating a microstructured 3D p-i-n junction with low reverse current. It permits to suggest, that made by PIII method p-n junction with a wide depletion area would be had leakages, acceptable for betavoltaic power source.

References

1. Kuan-Chih Huang, R. Dahala, James J.-Q. Lu, Y. Danon, I. B. Bhat. Continuous p-n junction with extremely low leakage current for microstructured solid-state neutron detector applications.// Proc. of SPIE Vol. 8710, 87101J (2013).

2. Kuan-Chih Huang, R. Dahala, James J.-Q. Lu, Y. Danon, I. B. Bhat. High detection efficiency micro-structured solid-state neutron detector with extremely low leakage current fabricated with continuous p-n junction. // APL, v. 102, 152107 (2013).

3. Kuan-Chih Huang, R. Dahala, James J.-Q. Lu, A. Weltz, Y. Danon, I. B. Bhat. Scalable large-area solid-state neutron detector with continuous p–n junction and extremely low leakage current. // Nuclear Instruments and Methods in Physics Research A763, p.260–265 (2014).

4. Handbook on Plasma Immersion Ion Implantation and Deposition.// Ed. by A. Anders /Wiley, 760 pp. (2000).

5. Shahsenov, I. S., Miakonkikh A.V. and Rudenko K.V. Monte // Carlo simulation of boron doping profile of fin and trench structures by plasma immersion ion implantation / Proc. SPIE 9440, International Conference on Micro- and Nano-Electronics 2014, 94400Y (December 18, 2014)

6. Chengkun Xu, Zhang Xi, King-Ning Tu et al. // Journal of the Electrochemical Society. 2007. V. 154(3). V. D170.

Planar betavoltaic converter creation with plasma-immersion ion implantation process*

K V Rudenko[1], A V Miakonkih[1], A E Rogojin[1], S V Bogdanov[2], V G Sidorov[2], P V Zelenkov[2]

[1] Institute of Physics and Technology of the Russian Academy of Sciences, Moscow, Russia

[2] Reshetnev Siberian State Aerospace University, Krasnoyarsk, Russia

E-mail: zelenkow@rambler.ru

Abstract. Some results on planar diode structure creation by the method of a plasma-immersion ion implantation is presented in this paper. Obtained leakage current ~ 1 uA/cm^2 at reverse voltage -1 V.

Keywords: betavoltaic element, p-i-n diode, plasma-immersion ion implantation.

Introduction

Direct conversion radioisotope micropower sources have drawn significant attention because of their potential of providing a long-term energy solution for low-power consuming microsystems. In recent years, a series of studies have been reported on betavoltaics with planar silicon p-i-n diodes [1-8], irridiated with Ni-63 isotope layer. The efficiency of these p-i-n structures rely on high-quality p-n junctions, in particular low leakage currents.

The similar research in the field of betavoltaics carrying out with efficiency enhancement of solid p-n junction neutron detectors [9, 10]. Common to these applications is the task of effective betaelectrons silicon detector creating, based on p-n junction with extremely small leakage current. In [10-12] shown, that planar epitaxial diode structures with narrow junction field region has leakage at the level ~1 uA/cm^2; same p-i-n diode with wide junction field region has leakage about 20 nA/cm^2.

In this research, authors research possibilities of application of a plasma-immersion ion implantation process (PIII) of boron into silicon were analyzed, also planar p-n junction was formed and leakage currents were measured to compare with parameters of epitaxial junction formed in [9].

Production and measurement of the planar p-n junctions

Silicon wafers «КЭФ-4,5 (100)» (n-type silicon wafers, doped with Phosphorus, 4,5 Ohm·cm volume resistivity) were used to experiment. Manufacturing mode is selected so that formed with PIII method p-n junctions were similar in concentration to epitaxial structures,

* Research is performed with support from Russian Federal Targeting Program «Research and development for priority targets of Russian scientific-technological complex development years 2014-2020», agreement 14.577.21.0117, unique identifier RFMEFI57714X0117

studied in [9]. The wafers passed preliminary high purity reagents cleaning with next scheme:

1. Isopropyl alcohol about 1 minute.
2. De-ionized water 5 min.
3. H2O2(50%):NH4OH(25%):H2O (1:2:5) 1 min.
4. De-ionized water 1 min.
5. HF (10%) 1 min.
6. De-ionized water 1 min.
7. Dry nitrogen continuous-flow drying

After preliminary cleaning, then was carried out surface layer doping by BF_2+ ions with the PIII method on a laboratory plant. The laboratory plant has low-pressure high frequency plasma ICP-source, in plasma-forming BF_3 gas, where base ion is BF_2^+. The ion implantation process passed in pulsing periodic mode, with applying rectangular accelerating negative polarity impulse to wafer. Impulse potential ($-V_{acc}$) define energy of the implanted ions; the relation between duration of constant potential to duration edge/drop of impulse define energy dispersion. Impulse duration was 10 us, the form is rectangular, pulse repetition frequency 1 kHz, amplitude 5 kV. Thus, energy of the implanted ions was 5 keV. The process of formation of the implanted structures include the following:

1. Preamorphisation of silicon facial layer by ions He^+ for channeling effect exceptions of implantable boron into silicon.
2. Large dose implantation of boron ($D = 2x10^{15}$ cm^{-2}) from gas plasma BF_3 in immersion mode by boron-containing molecular ions BF_2^+ (dominant), also with a small amount of BF+ and B+.
3. Exposure surfaces of the implanted layer in hydrogen plasma to eliminating of boron precipitates from a surface, formed by large-dose implantation process of low energy boron ions.
4. The stage of cleaning a surface from boron precipitates is compatible to the PIII process and it carried out in the same technological camera. The surface of the implanted structure is exhibited into hydrogen plasma without giving of electric shift on a plate at temperature of 300^0C.
5. After implantation process of annealing and electric activation of boron was carried out by fast photon annealing. The plates were annealed in the atmosphere Ar (99,9%) at temperature of 780 °C. Annealing time of admixture was 10-25 minutes, including a stage of the diffusive acceleration.
6. Laying tungsten on a silicon surface was carried out by the method of magnetron dispersion. Dispersion was carried out in DC mode in argon atmosphere, thickness of tungsten layer is 150 nm. The backward contact was created with analogous method.

Measurements of I-V characteristic in structures formed by PIII method (area about 1 cm^2) were carried out at the room temperature with probe station Micromanipulator 7000, which are connected to the system Keithley 4200-SCS.

First of all interest was caused by the reverse bias of the diode characteristic, that determines reverse current, and standard dependence is given in a figure 1.

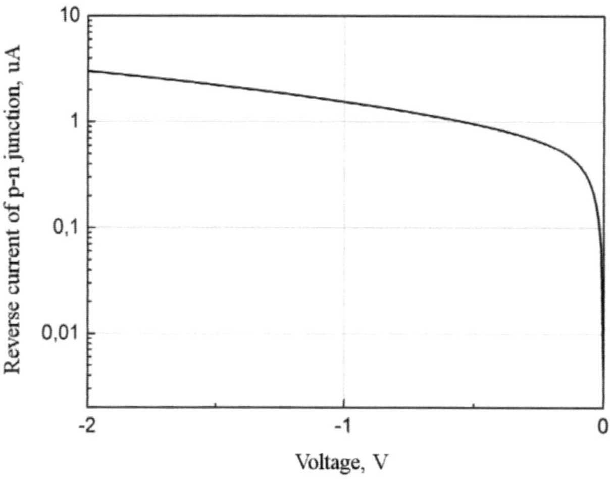

Figure 1. The reverse bias of the diode characteristic formed by the PIII method p-n junction which measures on mesa structure.

Conclusion

Thus, the results obtained in p-n planar diode structure, formed by plasma-immersion ion implantation process, by leakage are at the level of ~ 1 uA/cm2 at U = -1 V, which corresponds to the epitaxial planar junctions obtained by high-temperature CVD-methods. It permits to suggest, that made by PIII method p-n junction with a wide depletion area would be had leakages, acceptable for betavoltaic power source.

References

1. V.N. Murashev, V.N. Mordkovich, S.A. Legotin, et al., J. Nano- Electron. Phys. 6 No 4, 04012 (2014)

2. Murashev V.N., Legotin S.A., Didenko S.I. et al. // Advanced Materials Research 2015. V. 1070-1072. P. 585

3. A.A. Krasnov, S.A. Legotin, Yu.K. Omel'chenko, et al., J. Nano- Electron. Phys. 7 No 4, 04004 (2015)

4. S.Yu. Yurchuk, S.A. Legotin, J. Nano- Electron. Phys. 7 No 3, 03014 (2015)

5. S.U. Urchuk, A.A. Krasnov, S.A. Legotin, et al., J. Nano- Electron. Phys. 7 No 4, 04005 (2015)

6. A.S. Korolchenko, S.A. Legotin, V.N. Murashev, M.N. Orlova Metallurgist 54 No 5-6, 328 (2010).

7. A.S. Korol'chenko, S.A. Legotin, S.I. Didenko, S.P. Kobeleva, Russ. Microelectronics 40 No 8, 620 (2011).

8. V.N. Murashev, A.S. Korolchenko, S.A. Legotin, Metallurgist 56 No 3-4, 303 (2012)

9. Kuan-Chih Huang, R. Dahala, James J.-Q. Lu, Y. Danon, I. B. Bhat. Continuous p-n junction with extremely low leakage current for microstructured solid-state neutron detector applications.// Proc. of SPIE Vol. 8710, 87101J (2013).

10. Kuan-Chih Huang, R. Dahala, James J.-Q. Lu, Y. Danon, I. B. Bhat. High detection efficiency micro-structured solid-state neutron detector with extremely low leakage current fabricated with continuous p-n junction. // APL, v. 102, 152107 (2013).

10. Kuan-Chih Huang, R. Dahala, James J.-Q. Lu, A. Weltz, Y. Danon, I. B. Bhat. Scalable large-area solid-state neutron detector with continuous p–n junction and extremely low leakage current. // Nuclear Instruments and Methods in Physics Research A763, p.260–265 (2014).

11. Handbook on Plasma Immersion Ion Implantation and Deposition.// Ed. by A. Anders /Wiley, 760 pp. (2000).

12. Shahsenov, I. S., Miakonkikh A.V. and Rudenko K.V. Monte // Carlo simulation of boron doping profile of fin and trench structures by plasma immersion ion implantation / Proc. SPIE 9440, International Conference on Micro- and Nano-Electronics 2014, 94400Y (December 18, 2014)

IOP Conf. Series: Materials Science and Engineering **122** (2016) 012030 doi:10.1088/1757-899X/122/1/012030

Modern educational information technologies in formation of communicative competency in a foreign language in the process of training postgraduates at the aerospace university

M V Savelyeva and N A Shumakova

Reshetnev Siberian State Aerospace University, 31, "Krasnoyarskiy Rabochiy" prospect, Krasnoyarsk, 660037, Russia.

E-mail: mvsavelyeva12@gmail.com

Abstract. The research analyses the issues of competency-based approach implementation in connection with practical experience gained in SibSAU in teaching English for scientific purposes to postgraduate students. The article focuses on IT application for the objectives of both class room and independent work of post graduates organization.

It is the main function of education that is in general aimed at training useful professionals to different parts of life taking into account current tendencies and the level of science and technology the people should correspond to; consequently, the critical factor for further development of education is the choice of a leading educational approach. The teaching approach has to address the society needs such as contributing to process of globalisation and innovative development, which means educating and bringing up people capable of realising themselves under the determined conditions.

The education in Russia has experienced the serious changes. The discrepancy of the recent knowledge-based approach in modern life is its inability to meet the challenges: providing knowledge and skills about the world to modern generation is not enough as they aspire to realise their own preferences and interests within their professional performance. The knowledge-based approach is closely connected with teaching and instructing people, which does not allow them to unlock their creativity, independent critical thinking and develop the ability to find their places under constantly changing conditions. This causes a problem: the graduates of universities and colleges cannot easily meet the requirements of life, feel lack of both personal and professional experience and are unable to realise themselves. It results in frustration and dissatisfaction.

The knowledge-based approach develops intellect and memory, but motivation of students is not affected as well as their sphere of emotions and feelings. The approach greatly contributes to the development of formal level of education. The main shortcoming of the approach is that graduates feel hard to adapt for the professional life.

The current leader among existing approaches is competency-based, it is designed to train and to provide the development of necessary competencies for students based on their understanding the requirements they will meet as employees. Students comprehend what competencies they are interested in and an educator finds the methodologies and techniques to realise the training process. Training differs from teaching. The essence of training is to focus on practical skills more than pure theoretical knowledge and the ability to apply the skills more than demonstrate the theoretical awareness of the subject. [1]

The competency-based education approach found its popularity due to its application in the vocational training. European and Australian educational systems also used the approach to measure professional skills. [2]. Richards J.S., Rodgers T.S. point out the significant impact of competency-based approach on language teaching and the results obtained while teaching survival language to immigrants. [3]

During the training process first of all the students achieve the competencies they need to be successful employees that result makes them feel confident about their future. The competency-based approach allows educators to use training time more efficiently, that is rather significant within the time limits provided by the curriculum. The competency-based approach involves more individual assignments to the students and educators' possibility to feedback every student effort, thereby, students' and educators' cooperation is rather fruitful as they both know the objectives to reach.

However, certain limitations of the competency-based approach could interfere with educational process based on it. Due to the results obtained by the observation and analysis of 4 year competency-based approach used in teaching practice we could agree with Richard S. Sullivan [4], who presents in his study the findings of the limitations of the approach:

- the first problem to overcome is educators' position, who consider their own previous learning experience to be a model for further generations of students. Therefore, they continue applying those traditional teaching approaches they have been accustomed to;
- the second shortcoming is a logic continuation of the first. Educators' negative attitude to a competency-based approach can be reflected in the choice of teaching methodologies though the materials are designed according to competency-based approach principles.
- the third limitation is not relevant identification of the job competencies by the students that results in ineffective teaching process. The current difficulty is the students' poor motivation to prepare for the future job. They are not aware of the skills they have to develop to become professionals.

We analyse the students' attitude to the foreign language competencies every academic year [5], according to our survey the percentage of those who find it not so significant to acquire the competencies relevant to their future job ranges from 10 to 15 at bachelor courses and it reduces to 1.5% at master programmes, the post-graduate students have the highest motivation in developing proper competencies as scientists. We have chosen for the analysis the most motivated audience of postgraduates as the most motivated to develop competencies necessary to do research. Moreover, in SibSAU great experience of work with post graduates on foreign languages for scientific purposes is accumulated. Working with postgraduates in this field we use a complex approach. Further we'll dwell upon the main elements of this complex in more details. In particular these elements are practical and theoretical at different stages of the course. Special attention herein will be paid to the results and the forms of control.

The methodology of teaching English for scientific purposes is connected with effective planning and organizing both classroom activities and independent students' work. The academic content of English for scientific purposes course in this context is one of the most important elements of the integral structure of academic process planning. This process is aimed at formation of different universal competencies, such as 1) the ability to participate in the collaborative work of international and Russian scientific teams in solving both scientific and educational problems; 2) availability to use modern methods and techniques of scientific communication in a foreign language. In connection with this there is a need for analyzing pedagogical and methodological terms, contributing the efficiency of training.

As it is said, the objective of training is formation of universal and communicative competencies, which is achieved through the content of the programme, consisting of classroom activities and independent work.

Here it is important to underline the significance of independent work. It is divided into two units: work at distant learning server (Moodle platform) and search for authentic monograph in the field of research in English. Postgraduates perform various tasks placed by teachers of foreign languages at the distant learning server, e.g. watching video on the programme's topics and expressing their opinion after watching in writing, writing CV, glossary, annotated bibliography (formation of scientific writing skills). Besides they prepare different presentations.

Presentation has a special place during classes for post graduates. One of the topics is "My dissertation proposal" where they reinforce the skills on formulating the topic, object and subject of the research; present its main tasks in English. Another example of the presentation is "Peer-reviewed journals of Scopus and Web of Science systems in my field of research". Here the postgraduates present different journals in their field of study paying special attention to the terms of publishing (formation of speaking and presentation skills).

Public presentation of the research report is a final part of the course and assessment tool.

Underlining the importance of the work it is also significant to say about preparation to publish an article in English in the edited collection of the scientific conference in SibSAU. If we return to the importance of the result, we could speak about the following: presentation of the speeches at the scientific conference (public speaking competency), presentation of test translation of the monograph (scientific writing skill). Having analyzed different types of work, performed by the postgraduates, we can conclude that all the elements of the process and the result are interconnected and correspond to the main objective of competencies formation. Fig.1 demonstrates an increasing tendency of postgraduates' research activity due to training based on competency-based approach.

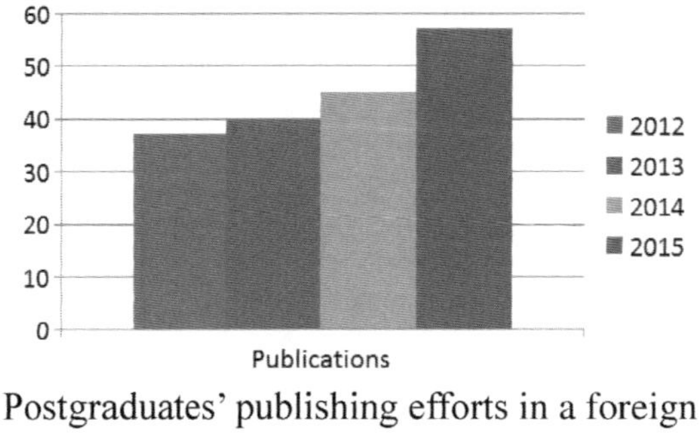

Figure1. Postgraduates' publishing efforts in a foreign language.

Internet resources have already become an integral part of the educational process in Sib SAU. These tools started to be actively used in 2006-2007. It should be noted that the use of the Internet greatly influences the technology of delivery information to the postgraduate students. In these terms the role of independent work is increasing and the autonomy of studying is developing.

The use of Internet for teaching English for scientific purposes amid the global economic crises becomes justified, as it reduces expenses on paying for supplementary education (foreign language courses), buying expensive authentic textbooks and monographs which get out of date and loose novelty very quickly. Besides, electronic publications provide achievement of four basic advantages: international accessibility, speed, additional facilities and low cost. [7]

Information and communication technologies create new global environment, mediating a great number of Internet resources, supplying academic and scientific information in foreign (English) language. In the methodology of teaching Foreign languages there exists a new direction such as e- learning having been developed both by foreign and national researchers. Internet resources enable to teach all types of speech activity, though teaching writing and reading are higher-priority ones.

Great number of Internet resources was investigated by the specialists of SibSAU. Some useful scientific search systems, catalogues and back files of scientific periodicals were revealed and could be recommended to the masters and postgraduates of various directions in their work with authentic textbooks, journals, monographs.

For technical students the following recourses could also be recommended:

https://tefltastic.wordpress.com/worksheets/technical-english/; [8]

http://eslrightnow.com/english-for-engineering/; [9]

http://www.youtube.com/channel/UCS5ME54d_iJ0cOOZun4ikKQhttp://www.proprofs.com/quiz-school/story.php?title=technicalengineering-english-test;

http://www.monash.edu.au/lls/llonline/grammar/index.xmlgrammar for engineers;[10]

http://www.cambridge.org/it/elt/catalogue/subject/project/custom/item2489163/Cambridge-English-for-Engineering-Resources/?site_locale=it_IT&currentSubjectID=2561588 case studies for Cambridge English for Engineering.

Besides we highly recommend the following resources for development of basic linguistic skills:

http://www.scribd.com/ is for development of reading skills, creation of individual materials and tests for reading in IELTS format and in other formats, giving the opportunity to share the texts and to see the ratings;

http://www.ted.com/, http://www.bbc.co.uk/worldservice/learningenglish/general/sixminute/ are examples of video and audio podcasts with the possibility of text creation;

http://azargrammar.com/index.html, http://www.englishpage.com/index.html are grammar and grammar tests;

http://voicethread.com/#home, http://englishcentral.com/home?ref=logo, http://vocaroo.com/ are links for development of speaking skills with the emphases on pronunciation and speech fluency;

http://www.listen-and-write.com/ are for development of listening and writing skills, https://docs.google.com/#home is the link for creation of Google form tests enabling to conduct collaborative editing and discussing.[11]

The level of language proficiency is an important criterion of aerospace industry specialists' professionalism as there is a necessity to deal with foreign colleagues, obtain information from international scientific sources, participate in international scientific events and to publish the results of their researches in the peer-reviewed journals with high citation index. The knowledge of English for scientific purposes enables young researchers to transfer the results of the performed research to the international level. Nowadays SibSAU researchers actively participate in international scientific and academic projects. Scientific contacts with our foreign colleagues have become numerous. All the above mentioned factors caused the necessity of English for scientific purposes skills development. In this case English for scientific purposes is considered as a means of obtaining adequate modern knowledge from scientific professional sphere. [6]

As a result of learning a course of English for scientific purposes post graduates should get the experience of large volume of authentic scientific texts (monographs, scientific articles, reports of international specialists) processing. They should be able to write abstracts for scientific articles and other research projects, grant proposals in order to participate in scientific conferences, to draw up application forms for conferences.

The analysis of each separate effective technology, used at different stages of teaching postgraduates is a complicated longtime process and can become the subject of a special discussion. In this connection one can mention that the implementation of Internet technologies into the teaching process of postgraduate students positively effects cognitive-informative, professionally-oriented and didactic potential. It creates positive emotional educational environment, developing autonomy, ability of critical thinking and independent decision making.

Having analyzed the process of teaching postgraduates of aerospace profile English for scientific purposes on a complex based methodology developed in SibSAU we can conclude that curriculum content is aimed at postgraduates' comprehensive training in the field of professional communication in the international educational and scientific environment. The result of our work is expressed through formation of universal and language competencies. Positive experience of integration of Internet resources, e-learning and other innovative forms of work in the process of training is accumulated. Effective methods of training are worked out.

References

[1] Uzun Levent. 2012. What is your educational philosophy? Modern and post modern approaches to foreign language education. SSLT **2(3)** pp.333-348

[2]Griffith W.I., Hye-Yeon Lim. 2014. Introduction to Competency-Based Language Teaching. MEXTESOL Journal. Vol. **38** No. 2. pp. 1-8

[3] Richards J.S., Rodgers T.S. 2001 Approaches and Methods in Language Teaching. 2nd edition. — Cambridge University Press. — 278 p.

[4] Richard S. Sullivan. 1995. The competency-based approach to training. U.S. Agency for International Development. JHPIEGO Strategy Paper No. – pp.1-9.

[5] Savelyeva M.V. 2015. Foreign language for post-graduate students: upgrading the course contents within the framework of the existing approaches. // IX International conference "Modern tendencies in foreign language teaching" – pp.20-25

[6] Shumakova N.A. Use of information technologies for implementation of integrated methods in teaching aerospace specialists English for scientific purposes. Vestnik Sibirskogo gosudarstvennogo aerokosmicheskogo universiteta imeni akademika M. F. Reshetneva (Vestnik SibGAU) , Vestnik SibSAU, Vol.16, № 3 , p.p.632-636.

[7] Vasil'eva P.A., Kochergina O.A.Integratsiya internet resursov v obuchenie inostrannomu yazyku studentov neyazykovogo vuza v usloviyakh global'nogo krizisa[Integration of internet resources in teaching foreign languages students of non-linguistic university in the conditions of global crises]. Cbornik nauchnykh statey "Global'ny ekonomicheskiy krizis: realii I puti preodoleniya [Collection of scientific articles "Global economic crises: reality and ways out"]. St. Petersburg, Institut bizneca I prava Publ., 2009, p. 30–32. (In Russ.)

[8] TEFLtastic blog.Available at:https://tefltastic.wordpress.com/worksheets/technical-english/(accessed 23.11.2015.)

[9] ESL Right Now.Available at:http://eslrightnow.com/english-for-engineering/(accessed 20.02.2015).

[10] Monash University. Available at:

http://www.monash.edu.au/lls/llonline/grammar/index.xmlgrammarforengineers (accessed 23.11.2015).
[11] Kuz'mina T.A. The experience of technologies and WEB2 instruments usage in teaching foreign languages [Opyt ispol'zovaniya tekhnologiy i instrumentov WEB 2 v prepodavanii akademicheskogo angliyskogo(ESP/EAP) v NIU VShE]. Materialy ezhegodnoy mezhdunarodnoy konferentsii "Professional'no orientirovannoe obuchenie inostrannomu yazyku I perevodu v vuze: materialy ezhegodnoy mezhdunarodnoy konferentsii [Materials of annual international conference "Professional'no orientirovannoe obuchenie inostrannomu yazyku i perevodu v vuze]. Moscow, 2013, p.165–166. (In Russ.)

IOP Conf. Series: Materials Science and Engineering **122** (2016) 012031 doi:10.1088/1757-899X/122/1/012031

Development of a model representing systems protected against research*

M Styugin
Research fellow, Siberian State Aerospace University named after Academician M.F. Reshetnev, Krasnoyarsk, Russia

E-mail: styugin@gmail.com

Abstract. The research described herein presents a method of modeling information systems with dynamic parameters. The model includes elements with discrete internal states and agents, which can act in the system by modifying states of the system's elements. The research has two objectives aimed at protection against research, which are as follows: inadmissibility for the system to transfer into some specific states and invisibility of actions taken by the agents. The present paper establishes stable secure states and proves algorithmic insolubility of Problem 1 in its general case..

1. Introduction

The task of information security lately is increasingly turning to the area of system protection against research by intruders at reconnaissance stage. The most well-known method in that area is the Moving Target Defense [1]. Over 150 different MTD methods [2] appeared within previous 5 years. However, problems related to evaluating efficiency of such methods still exist. Research studies [3, 4] are almost the only ones in this field. The present paper continues research started in [5] and [6]. The system will be represented as a unit, which consists of independent elements and agents. Each one of the agents is able only to see restricted system's states..

2. Developing a functional model

Assume a system could be represented as a set of its states T. The system can transform from one state to another as a result of actions of agents.

Let us number agents in the system 1 to *n*. Each of the agents has a set of actions F_k. Each action represents a system mapping function from one state to another one:

$$\forall F_i \in \mathbf{F_k} \quad F_i : T_{i-1} \to T_i$$

Hence, we can regard the composition of actions as a sequential transition from one state to another:

$$F_i \circ \ldots \circ F_s : T_{i-1} \to T_s$$

* The research was done with the financial support of the Ministry of Education and Science of the Russian Federation in accordance with the agreement № 14.574.21.0126, unique identifier RFMEFI57414X0126

Thus actions or their combination are equal when initial and final system states are equal.

$$F_i \circ \ldots \circ F_s : T_{i-1} \to T_s,\ F_k : T_{k-1} \to T_k,$$
$$F_i \circ \ldots \circ F_s = F_k \Leftrightarrow T_{i-1} = T_{k-1} \,\&\, T_s = T_k.$$

To show the information component of the conflict it shall be noted that each of the agents sees the system's set of states differently and observes its changes differently. The reflexive structure or the awareness structure of agents in the system here could have been represented as a sequence of T sets in the conception about each other, e.g. as set forth in [7] or [8]. Although such representation of awareness is a most general one, but it is of little use from the practical point of view as in its case it is unclear where to obtain data for reflexive structure or awareness structure.

The present paper uses an assumption that not every system element is equally well observed by all agents. Hence, some particular system's states are indiscernible for some agents. Such indiscernibility may be represented as the equivalence relation on set T.

$$\forall k = 1..n \ \ \exists R_k \in \mathbf{R}$$

Each system's state is represented as an aggregate of independent elements

$$T_i = \left(a_i^1, a_i^2, \ldots, a_i^n\right), a_i^1 \in \mathbf{A}^1, a_i^2 \in \mathbf{A}^2, \ldots, a_i^n \in \mathbf{A}^n$$

Two system's states T_i and T_j are indiscernible for agent k when

$$T_i R_k T_j \ \text{or} \ \left(a_i^1, a_i^2, \ldots, a_i^n\right) R_k \left(a_j^1, a_j^2, \ldots, a_j^n\right).$$

When in that case one of the elements is $a_i^l \neq a_j^l$ hence element a^l is indiscernible for agent k in T system.

It is also evident that the agent cannot take actions resulting in changes of states in elements that are indiscernible for the agent. For instance the situation below is impossible for agent k:

$$F_2 \in \mathbf{F_k} \quad F_2 : T_1 \to T_2,\ T_1 = \left(a_1^1, a_1^2\right), T_2 = \left(a_2^1, a_2^2\right),$$
$$\left(a_1^1, a_2^2\right) R_k \left(a_2^1, a_2^2\right) \,\&\, a_1^1 \neq a_2^1.$$

In the example above the first system's element changed its state as a result of agent k's actions despite that its state is indiscernible for k.

Thus it may be concluded that when a system changes its state due to actions taken by agent k

elements indiscernible for the agent remain unchanged.

$$F_j \in \mathbf{F_k} \quad F_j : T_i \rightarrow T_j .$$
$$\forall a_i^s \in T_i, \forall a^s \in \mathbf{A^s}, s = 1..n, (\ldots, a_i^s, \ldots) R_k (\ldots, a^s, \ldots) \Rightarrow a_i^s = a_j^s .$$

The above law may be denoted as the rule of indiscernible element conservation.

Consequence 1. An agent cannot transform a system into states indiscernible for that agent in accordance with its equivalence relations.

Undesirable states need to be introduced into system T so that the above model would allow formulation of the problem of system protection against research. Set of such states shall be noted as T^N.

$$\mathbf{T^N} \subset \mathbf{T}$$

All other states of the system may be regarded as acceptable.

$$\mathbf{T^P} = \mathbf{T} \setminus \mathbf{T^N}$$

May be represented by a tuple:

$$\left\langle \mathbf{T}, \mathbf{T^N}, \mathbf{R}, \{\mathbf{F}\}, T_0 \right\rangle$$

Where T_0 is the current system state and {F} is a set of sets for all actions of agents in the system.

$$\{\mathbf{F}\} = \{\mathbf{F_1}, \mathbf{F_2}, \ldots, \mathbf{F_n}\}$$

Based on the above model two different research protection tasks can be formulated:

1. The first task is related to preventing the system from transition to a state belonging to the unacceptable set T^N. This problem can be solved from the research protection point of view only by correcting indiscernible system elements in relation to agents.

When state T^N is reachable in that system then it shall be denoted as a function's value:

$$\left\| \left\langle \mathbf{T}, \mathbf{T^N}, \mathbf{R}, \{\mathbf{F}\}, T_0 \right\rangle \right\| = 1$$

Hence when T^N is not reachable, then

$$\left\| \left\langle \mathbf{T}, \mathbf{T}^{\mathrm{N}}, \mathbf{R}, \{\mathbf{F}\}, T_0 \right\rangle \right\| = 0$$

2. The second task is related to the difficulty of recognizing an agent which brought the system to the state belonging to T^{N} set or to any state in general. It is possible to detect such an agent with the present model by finding the chain of actions in its set which leads to an unacceptable state.

3. Solution to the first problem

First let us review a system with one agent:

$$\left\langle \mathbf{T}, \mathbf{T}^{\mathrm{N}}, R_1, \mathbf{F}_1, T_0 \right\rangle$$

Theorem 1. When $\forall T_j \in \mathbf{T}^{\mathrm{N}}\ \exists T_i \in \mathbf{T} \setminus \mathbf{T}^{\mathrm{N}} : T_i R_1 T_j$ and the condition $\forall T_j \in \mathbf{T}^{\mathrm{N}}\ (a_0^n,...,a_0^m) \not\subset T_j$ is fulfilled, where $(a_0^n,...,a_0^m)$ are such that $\exists (a_i^n,...,a_i^m) \subset T_i$ and $\exists (a_j^n,...,a_j^m) \subset T_j : a_i^n \neq a_j^n,...,a_i^m \neq a_j^m$ then $\left\| \left\langle \mathbf{T}, \mathbf{T}^{\mathrm{N}}, R_1, \mathbf{F}_1, T_0 \right\rangle \right\| = 0$.

Theorem 1 can be formulated differently as follows: when each system's unacceptable state has a corresponding acceptable indiscernible state for the only agent then the system will never transit to it provided that the system has not been in that state initially in accordance with indiscernible elements.

A system with several agents may be reviewed in a similar way:

$$\left\langle \mathbf{T}, \mathbf{T}^{\mathrm{N}}, \mathbf{R}, \{\mathbf{F}\}, T_0 \right\rangle.$$

Theorem 2. When $\forall T_j \in \mathbf{T}^{\mathrm{N}}\ \forall R_k \in \mathbf{R}\ \exists T_i \in \mathbf{T} \setminus \mathbf{T}^{\mathrm{N}} : T_i R_k T_j$ and condition $\forall T_j \in \mathbf{T}^{\mathrm{N}}\ (a_0^n,...,a_0^m) \not\subset T_j$ is fulfilled, where $(a_0^n,...,a_0^m) = \bigcap_{s=1..n} (a_0^g,...,a_0^h)_s$ and $(a^g,...,a^h)_s$ are such that for every s agent $\exists (a_i^g,...,a_i^h) \subset T_i$ and $\exists (a_j^g,...,a_j^h) \subset T_j : a_i^g \neq a_j^g,...,a_i^h \neq a_j^h$ is fulfilled, then $\left\| \left\langle \mathbf{T}, \mathbf{T}^{\mathrm{N}}, \mathbf{R}, \mathbf{F}_1, T_0 \right\rangle \right\| = 0$.

In case the problem settings of Theorem 1 and 2 are not fulfilled it does not imply that the system may transit to an unacceptable state. When all potential actions of agents are reviewed in accordance with the {F} set then chains of actions equivalent to action $F : T_o \to T_i, T_i \in \mathbf{T}^{\mathbf{N}}$. Let us consider the possibility to find the value of the characteristic function

$$\left\| \left\langle \mathbf{T}, \mathbf{T}^{\mathrm{N}}, \mathbf{R}, \mathbf{F}_1, T_0 \right\rangle \right\| \to \{0, 1\}$$

First it is required to find out whether there is an algorithm for defining the characteristic function and hence whether the problem is possible to be solved algorithmically.

Theorem 3. Finding the algorithm for calculating the system's characteristic function $\langle \mathbf{T}, \mathbf{T}^{\mathrm{N}}, \mathbf{R}, \mathbf{F_1}, T_0 \rangle$, is an undesidable problem.

The above theorems are examples of defining the Turing machine break points in a specific case. Given the restrictions applied to a specific task that list may be extended.

4. Solution of the second problem

The second problem is about the ability to correlate a transition in the system with an action of one of the agents. Here is a system of three agents and action F. Action F does not belong to a set of actions of any of the agents. However while a system transition took place hence action A may be represented as a composition of agents' actions. Assume that

$$F = F_1 \circ F_2 \circ F_3, \quad F_1 \in \mathbf{F}_3, F_2 \in \mathbf{F_2}, F_3 \in \mathbf{F_1}.$$

Definition 1. Trivial action is an action that cannot be presented as a composition of other actions which do not incorporate the action in question.

Consequence 2. An action that is not included into action sets of other system's agents cannot be trivial.

Definition 2. A simple composition is a conception of an action that cannot be written with less elements. That is when there is conception $F = F_4 \circ F_5$ hence $F = F_1 \circ F_2 \circ F_3$ is not a simple composition.

Assume that in our case the chain $F = F_1 \circ F_2 \circ F_3$ is a simple one. It may be concluded that transition F in the system may be performed only by combined actions of agents 1, 2 and 3. Nevertheless it is not so, because each of the above actions may be not trivial. The whole composition of action is expanded as shown on Fig.1. Agent numbers are indicated above each action accordingly to an agent's set of actions in which they are located.

That is an action F_1 may be presented as a composition of $F_{11} \circ F_{12}$, where где $F_{11} \in \mathbf{F_2}$, $F_{12} \in \mathbf{F_1}$. Action F_{11} may be presented as a composition $F_{111} \circ F_{112}$, where $F_{111} \in \mathbf{F_1}$ & $F_{111} \in \mathbf{F_2}$, $F_{112} \in \mathbf{F_1}$.

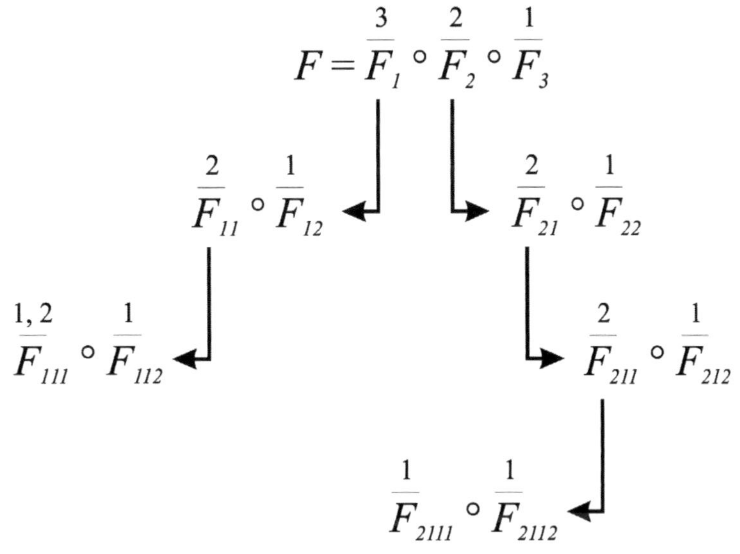

Figure 1. Expansion of a simple chain of nontrivial actions

As a result action F can be presented as a composition:

$$F = ((F_{111} \circ F_{112}) \circ F_{12}) \circ (((F_{2111} \circ F_{2112}) \circ F_{212}) \circ F_{22}) \circ F_3 .$$

All actions of the above composition are members of F_1 set. Hence agent 1 may have performed *F* transition on its own.

When an action in a simple chain is replaced with a simple chain corresponding to it then it shall be referred to as a simple composition of a second order. When some action in the second chain is represented as a simple chain then it is a simple composition of third order. The above composition is a fourth order composition.

When no limitations are applied to representation of actions then in a general case all possible simple compositions in the system (functional decomposition) can be found only by enumeration.

Conclusion 3. The objective of functional decomposition has an exponential complexity from $|\{\mathbf{F}\}|$

When set F is big enough it is impossible to perform functional decomposition hence it is impossible to find the agent which performed a transition in the system.

All this time the system was analyzed only from the outside with absolutely full awareness of its transitions and sets. In such conditions the task is always accomplished provided possession of unlimited computational power. From the system agents' point of view the task cannot be accomplished due to indiscernibility of agents in some elements. Thus a transition in the system performed using functional decomposition may be unobservable for some agents.

Hence the system may appear in some state absolutely "unexpectedly" for agents. A most suitable definition for that problem is a term of "functional steganography".

The subject of functional steganography is finding transitions in the system that are represented by functional decomposition of one agent's actions and indiscernible for the other system's agents.

Hence there is a subset of actions from the area of functional steganography for each agent:

$$\forall i \in N \ \exists \mathbf{F}_i^{(st)} \subset \mathbf{F}_i \ .$$

That set might as well be empty. Transition from that set of actions means that other agents will associate that action (due to indiscernibility) with actions of the other system's agents:

$$\forall F_k \in \mathbf{F}_i^{(st)} \ \forall j \in N \setminus \{i\} \ \exists F_s \notin \mathbf{F}_i \ : F_k R_j F_s$$

Rather interesting practical problems may be solved by defining such sets of actions for the system's agents. An example may be a task of concealing actions of a security administrator in a system. The conclusion below may be useful fir that task:

Conclusion 4. When one of the system's agents has access to all system's elements and the other agents have a set of indiscernible elements then a set of actions in the functional steganography of such agent is never empty.

A most simple implementation of that method can be performed with access control policies. Discretionary security policy can be an example where all objects are initially independent and indiscernibility may be regarded as no {read} right in the object access control list for subjects.

Conclusion

The present paper analyzed a mathematical model of an information system, which allows modeling insufficient awareness of agents about some of the system's elements. The model can be used to find out, which system elements that are protected against research allow preserving the system in a secure state that is to avoid transition to an unacceptable state. The paper establishes initial states of an information system when it remains protected at all times. Two theorems were proven and based on the above accomplishment. Solutions obtained after resolving the problem provide an opportunity to design information systems protected against research with a proven security at the level of separate independent elements.

References

[1] Carvalho M., Ford R. Moving-target defenses for computer networks. // IEEE Security and Privacy. 2014. 12(2) pp. 73–76.

[2] JafarHaadiJafarian Q.D.,Al-Shaer E. Openflow random host mutation: Transparent moving target defense using software-defined networking // Proceedings of the 1st Workshop on Hot Topics in Software Defined Networking (HotSDN). New York: ACM, 2012. pp. 127–132.

[3] Jajodia et al. Moving Target Defense. Creating Asymmetric Uncertainty for Cyber Threats. Series: Advances in Information Security. London: Springer, 2011. 184 p.

[4] Jajodia et al. Moving Target Defense II. Application of Game Theory and Adversarial Modeling. Series: Advances in Information Security. London: Springer, 2013. 203 p.

[5] Styugin M. Protection against system research // Cybernetics and Systems. 2014. 45(4). pp. 362–372.

[6] Styugin M. The New Method of Security Development for Web Services Based on Moving Target Defense (MTD) Technologies // Proceedings of the International Conference on Network Security and Communication Engineering (NSCE2014). Hong Kong: CRC Press, 2014. P. 130–136

[7] Lefebvre V. A. The Structure of Awareness. Beverly Hills: Sage, 1977. 177 p.

[8] Novikov D., Chkhartishvili A. Reflexion and Control: Mathematical Models. London: CRC Press, 2014. 298 p.

IOP Conf. Series: Materials Science and Engineering **122** (2016) 012032 doi:10.1088/1757-899X/122/1/012032

The internet based on presence system technology*

M Styugin[1], **A Kaygorodov**[2]
[1]Research fellow, Siberian State Aerospace University named after Academician M.F. Reshetnev, Krasnoyarsk, Russia
[2]CEO, LLC SilverLab, Krasnoyarsk, Russia

E-mail: styugin@gmail.com

Abstract. In our study we analyze how to create the systems based on "communication-resource-presence"- technology. We formulate functional and architectural requirements. It is shown some new features in this systems that pertain to communication and information search on the Internet. The Internet passed three infrastructure stages from the communication between two hosts to the resource intermediation and communication in real presence systems. The systems based on the presence technologies have just started to develop. Our study shows what criteria's they must meet. One of this criteria is the division of site resources into the "rooms" logically separated from each other. The users can see and connect to each other. The contextual data of user presence in a particular "room" can be used when searching for them in the context of professional competence.

1. Introduction

The Internet was created as an instrument of communication and remains so today. However, it went through several basic stages. Surely the information transmission rate has increased, the Web design technologies have changed and a wide variety of protocols have appeared, but the development of the Internet is structured only in several basic stages.

Simple data transfer protocols that enabled two network hosts to exchange messages could be considered to be the first stage of development of the Internet [1]. At the same time, we need not regard particular technological solutions of setting up tunnels, it means an e-mail or the Skype videoconference, because all this refers to the means of communication. We call this systems C- systems (from the word «communication»). They are displayed by the symbols A, B, C and D on the fig. 1.

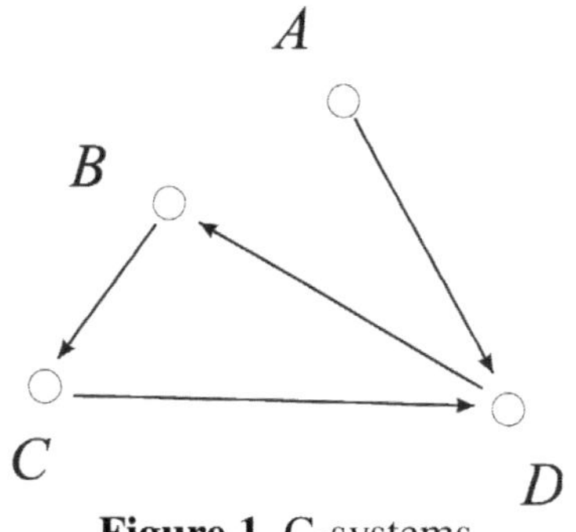

Figure 1. C-systems

* The research was done with the financial support of the Ministry of Education and Science of the Russian Federation in accordance with the agreement № 14.574.21.0126, unique identifier RFMEFI57414X0126

The second stage of development of the Internet is appearance of resources initially presented as Web pages using HTTP protocol. It does not make any difference whether this pages are static or dynamic, programmatically generated according to any variable values transferred via HTTP protocol. We call all this systems R-systems (from the word «resource»).

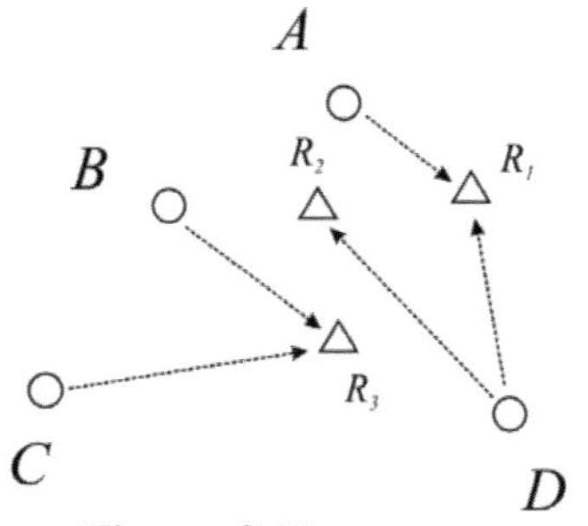

Figure 2. R-systems

Source hosts are displayed by the symbols R_1, R_2 and R_3 on the fig. 2. They are available in a wide area network according to the unique values of IP-address and return any information when requesting them.
The age of social networks and Web 2.0 created a special class of CR-system [2, 3].

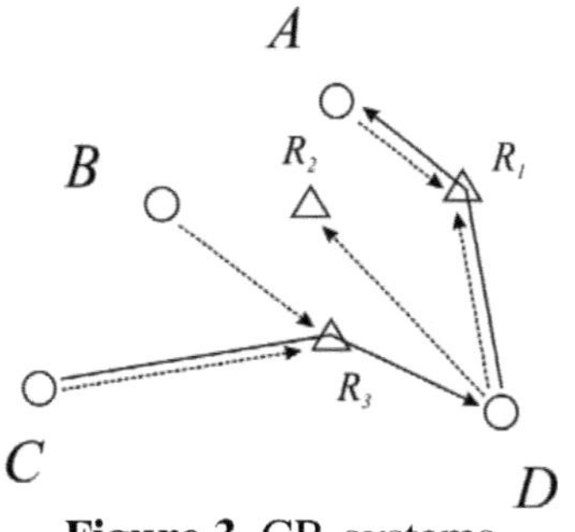

Figure 3. CR-systems

CR-system includes a possibility to communicate via Internet sources themselves. The online forum on the Web site is a classic example of the CR-systems that evolved into the multifunctional social networks.
The social networks were subsequently the first sources offered the option to fix whether the user is online or not. This was possible by registering the time of the last request to the site after a timeout at a given time interval. The fixation of user presence on the Web site at a particular time decided the appearance of P-systems (from the word «presence»).

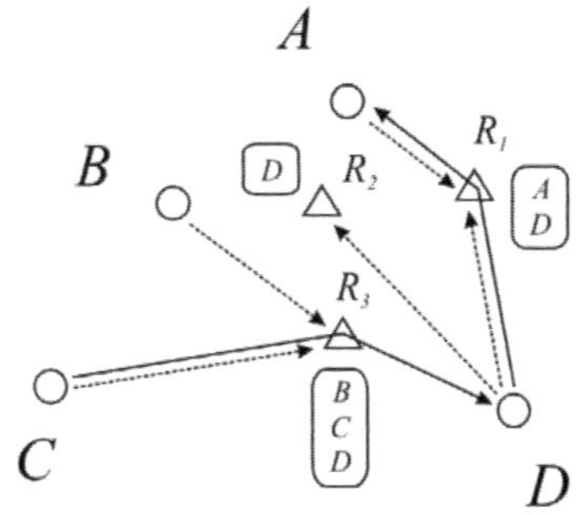

Figure 4. CRP-systems

The CRP-systems include not only CRP-system functional but also possibility to fix user presence at the same site where you are yourself.

There is little *doubt* that CRP-systems should be considered as separate stage of development of the Internet. But firstly, we should note – today there is no real Internet CRP-system that will be described in what follows. Secondly, the analysis of the CRP-system model, as we shall see, makes possibly to create new and more efficient technologies of communication and information search, the main functions of the Internet.

2. Structural model of the CRP-system

It is sufficient for our purpose to consider general structural pattern using RP-system as an example. The existence of an user authentication server is obligatory for the system because it is necessary for user identification on the third-party sites. In addition, the RP-Part module working on each page of Internet applications should be added for user identification.

The RP-Part module fixes the «room» of user presence and transmits the data to the server. The server transmits (according to the data on presence of other users in the «room») the data to the RP-Client that shows it to user. The RP-Client could be designed as a single console application, browser plug-in or could be built in at the RP-Part module.

The RP-Part module includes not only data on presence of users but also Room-filter that ranks the territory of user presence. The basic pattern of the Room-filter prescribes reserving of a separated «room» for each Web page (unique URL).

The site owner himself is expected to differentiate site rooms easily.

The difference between CRP-system and RP-system is in addition of a channel of direct communication between users through the CRP-Client that could be regular channel of short messages (messenger) or a videoconference.

3. Communication and information search in the CRP-system

CRP-system contains full information about sites visited by user, as well as all details of the text messages sent and received by user through the messenger.

When visiting Web sites, CRP-system enables to observe who is on the same site, who is reading the same article or is in the same forum thread as you are at the moment. It means the CRP-system gives a possibility to create threads in the context of presence both with single user and all users in the «room».

The second important advantage of communication in the CRP-system is that it enables to form a database that could be used for context search of conversation partner (with the same topics, requests etc.). It forms the basis of new and more efficient Internet communication technologies.

When searching for information, CRP-system enables to use the user presence data to improve the pertinency of queries. Today's search engines reached the limit of page relevance determination on the basis of their content and reference information. User, when looking through the pages from the search results, could better determinate relevance of pages for his query. This information could be used during the fixing of the «rooms» visited by user. For example, on the fig. 5 we can see that *Room₃* turned out to be the most relevant for the user.

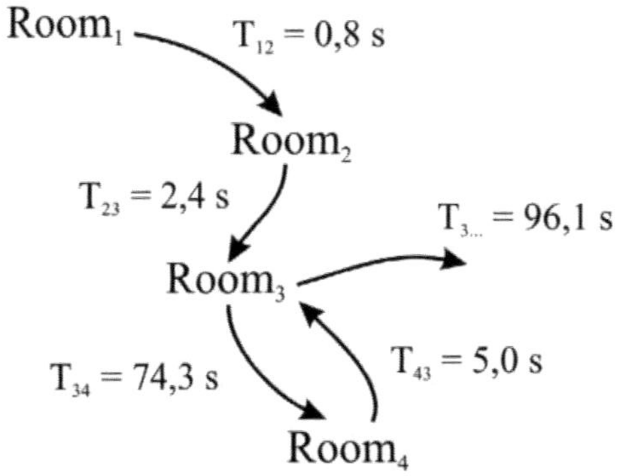

Figure 5. The pattern of user's motions through the «rooms»

4. Schedule drawing of the CRP-system structure

The process of the system operation is formed in the following manner: the particular programme code provided by CRP-system designer is embedded into the pages of Web sites. From now on, sites are included into the CRP-system and the site visitor monitoring process starts. The chat client is integrated into the interface of pages of the site involved. Site visitors who are on the same page could see each other on the list of online chat partners and could communicate with each other, they are in the same «room».

The whole CRP-system operation process consists of following:

- identification of site visitor;
- fixation of site visiting;
- providing a possibility of conversation with visitors who are on the same page.

Architecture of the system. The system needs particular software to operate properly according to the desired functionality.

User software:

- programme code embedded in the page;
- messenger.

Server software:

- server applications handling requests from sites; requests represent what pages user visited.
- server applications serving messenger, sending and receiving messages and doing other operations.

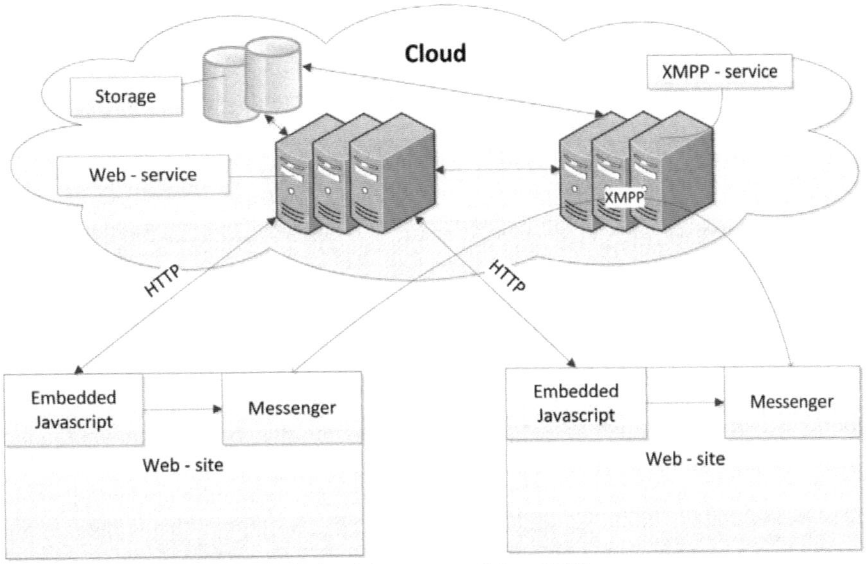

Figure 6. Architecture of the CRP-system

5. Prospects of further development of the CRP-system.

Creation of statistical and analytical services

During the operation, CRP-system accumulates large amounts of statistics. The most important data is:

- site traffic data;
- user time spent online;
- whether user communicated with each other and how the exchange of messages was intensive.

All this data could be available for site owners who use CRP-system.

The behavior of users on the site pages could be analyzed using statistics. As a result, the site owner could receive information about what pages the visitors are interested in, what pages cause discussions etc.

Social services. Using analysed data, it is possible for each user to form a «cloud» of his favorite topics and to choose for him suitable conversation partners. You could create a «topical portrait» of every user in his profile, for example, it could contain for one user world news, politics, business, public relations and for another – web server, algorithms, databases. The CRP-system users could form a social network with an emphasis on conversation where the friends would be «conversation partners» and there would be «topics for conversation» instead of groups. The news would report the hottest topics and reflect pages (that use CRP-system) with heated arguments and discussions.

Expansion of functionality of sites used with CRP-system. It is possible in the future that some solutions enable to integrate communication system of the site with the CRP-system. CRP-system could provide API that helps us integrate the communication system. The site is given following advantages:

- the site is given complete powerful communication system, there is no necessity to develop new own communication system;
- users could communicate with visitors of other sites and thus new visitors could be attracted from other sites;

IOP Conf. Series: Materials Science and Engineering **122** (2016) 012032 doi:10.1088/1757-899X/122/1/012032

- users of communication system could communicate using video and audio connection; it is difficult for common sites to provide such option.

Conclusion

The CRP-systems described in our study create new directions for functional development of Internet communication tools, as well as more efficient systems of information search and evaluation. The CRP-technology is considered to be an universal means to expanse functionality of Web sites based on HTTP-technology without using of Internet application design technologies.

References

[1] Marshall T.P., 2010, *A History of Communications: Media and Society from the Evolution of Speech to the Internet.* 1st ed. New York: Cambridge University Press.
[2] O'Reilly, T., 2005. *What Is Web 2.0. Design Patterns and Business Models for the Next Generation of Software* [online]. Available from: http://oreilly.com/pub/a/web2/archive/what-is-web-20.html.
[3] Chadwick A., Howard P.N., 2008, Web 2.0 and the Transformation of News and Journalism: New Possibilities and Challenges in the Internet Age. In: P N. Howard, eds. The Handbook of Internet Politics. New York: Routledge, pp. 218-233.

The heterogeneous anti-radiation shield for spacecraft*

S V Telegin, O N Draganyuk

M. F. Reshetnev Siberian State Aerospace University
31, Krasnoyarsky Rabochy Pr., Krasnoyarsk, 660037, Russia

E-mail: telegin@sibsau.ru

Abstract. The paper deals with modeling of elemental composition and properties of heterogeneous layers in multilayered shields to protect spacecraft onboard equipment from radiation emitted by the natural Earth's radiation belt. This radiation causes malfunctioning of semiconductor elements in electronic equipment and may result in a failure of the spacecraft as a whole. We consider four different shield designs and compare them to the most conventional radiation-protective material for spacecraft - aluminum. Out of light and heavy chemical elements we chose the materials with high reaction cross sections and low density. The mass attenuation coefficient of boron-containing compounds is 20% higher than that of aluminum. Heterogeneous shields consist of three layers: a glass cloth, borated material, and nickel. With a protective shield containing heavy metal the output bremsstrahlung can be reduced. The amount of gamma rays that succeed to penetrate the shield is 4 times less compared to aluminum. The shields under study have the thicknesses of 5.95 and 6.2 mm. A comparative analysis of homogeneous and multilayered protective coatings of the same chemical composition has been performed. A heterogeneous protective shield has been found to be advantageous in weight and shielding properties over its homogeneous counterparts and aluminum. The dose characteristics and transmittance were calculated by the Monte Carlo method. The results of our study lead us to conclude that a three-layer boron carbide shield provides the most effective protection from radiation. This shield ensures twice as low absorbed dose and 4 times less the number of penetrated gamma-ray photons compared to its aluminum analogue. Moreover, a heterogeneous shield will have a weight 10% lighter than aluminum, with the same attenuation coefficient of the electron flux. Such heterogeneous shields can be used to protect spacecraft launched to geostationary orbit. Furthermore, a protective boron-containing and nickel coating can be deposited onto a finished housing frame of space equipment.

Introduction.

When on orbit, satellite onboard electronic units require protection from space radiation causing malfunctioning of the spacecraft (SC) navigation and telecommunication systems [1-

* This work was supported by the Ministry of Education and Science of the Russian Federation, Grant No 169/14.

2]. A subsection of the Federal State Program "Space apparatus for fundamental space research" provides for elaboration of "space complexes and instruments to investigate Phobos, Mars, Venus, the Moon". European countries and the USA and China have intensified their activity on preparation for manned flights to the Moon, Mars and subsequent exploration and exploitation of these cosmic bodies. Modeling the effect of ionizing radiation on various materials is one of the important preparation stages aimed at development of new compositions for protective shields [3-8].

Most often protection is provided by the aluminum or aluminum-alloy housing of an electronic unit. The challenging research task in spacecraft protection from ionizing radiation is to come up with a shield that would offer better protective qualities and weigh less than its aluminum counterpart [9]. Properties of the materials are analyzed based on the following characteristics: the absorbed doze, mass thickness, mass attenuation coefficient, and the number of penetrated particles.

Experimental procedure.

Heterogeneous shields made of materials capable of absorbing or scattering hazardous radiation prove to be the most effective protective means to ensure reliable functioning of the control system of a satellite [10]. The ionizing radiation of an electron flow emitted by natural radiation belts of the Earth is the most dangerous on a geostationary orbit [11]. The number of solar and galaxy particles is grossly less, therefore modeling radiation as an electron beam with the energies of 0.04-5 MeV yields plausible results.

The choice of materials should rely on their physical properties, the reaction cross-section being one of them [12]. The higher is the atomic number of a chemical element, the more severe is the loss of energy by primary electrons; however, increasing the thickness induces brake radiation in heavy-element materials. So, it would be reasonable to consider using elements of different groups when modeling a protective shield [13]. A heterogeneous shield with layers of different density offers advantages that will be illustrated for the two models discussed below.

The protective shield consists of three layers (Table 1) arranged according to the growing atomic number of the basic absorbing substance of the layer. The first layer is a boron compound (boron carbide and sodium tetraborate shields have been considered). The second (structural) layer is a T-10 glass fabric containing aluminum, silicon and boron oxides and having a higher attenuation coefficient than aluminum. The last layer is nickel, having the highest atomic number. The choice of these materials is due to the reaction cross-sections in these media. Boron has the largest photon trapping - ($0.0092 \cdot 10^{-24}$ cm^2) and scattering ($3.6 \cdot 10^{-24}$ cm^2) cross-sections among the inexpensive, light and commercially available materials [12]. Nickel possesses the best characteristics such as $4.43 \cdot 10^{-24}$ cm^2 trapping cross-section and $17.3 \cdot 10^{-24}$ cm^2 scattering cross-section, which is an order of magnitude higher than the reaction cross-sections of other heavy elements, for instance, copper, their density being about the same. So nickel is a good material to protect against gamma-radiation and therefore it is efficient as the last protective layer [14].

Table 1. Protective shield structure

Shield/Layer	2.75/2.5 mm	1.7 mm	1.75 mm
1	B_4C	Glass fabric	Ni
2	$B_4Na_2O_7$	Glass fabric	Ni

Thicknesses of the layers obey the half-attenuation rule and are 2.75/2.5 mm for the boron-containing coating, 1.75 mm for the nickel layer, and 1.7 mm for the glass fabric.

Epoxy resin, which is 20-30% of the shield, was chosen as a binder.

Results and discussion.

The shielding qualities of the materials under study are summarized in Table 2. The data for nickel and boron-containing substances take into account the weight content of epoxy resin (25% in boron and 20% in nickel layers). The number of particles having penetrated the shield is normalized to one particle.

The mass attenuation coefficient decreases with growing density of the material and increases as the number of penetrated particles goes down. Boron-containing materials and glass fabric have a higher electron flow mass attenuation coefficient compared to aluminum, which supports the preferred choice of these materials for protective shields. Although nickel is not so good in interacting with electrons, it is advantageous in other aspects, which will be discussed further down.

A comparative analysis of the materials has also revealed that a homogeneous shield with a chemical composition identical to a heterogeneous shield has a 20% higher mass attenuation coefficient compared to aluminum.

Table 2. Comparing properties of the materials

Material	Density, kg/m^3	Thickness, m	Mass thickness, kg/m^2	Amount of penetrated electrons	Mass attenuation coefficient, m^2/kg
$B_4Na_2O_7$	1904	0.001	1.90	0.0069	2.6088
B_4C	1881	0.001	1.88	0.0079	2.5684
T-10	1596	0.001	1.60	0.0071	3.0270
Ni	3898	0.001	3.90	0.0006	1.9044
Al	2700	0.001	2.70	0.0031	2.1384
Homogeneous shield with $B_4Na_2O_7$	2156.4	0.001	2.16	0.0044	2.5146
Homogeneous shield with B_4C	2134.4	0.001	2.13	0.0032	2.6873

To understand the contribution of each layer to the radiation attenuation, we studied the amount of penetrated particles (electrons) depending on the shield thickness (Fig.1). The angle of inclination of the sections corresponding to different layers being about the same is an indication of the proper choice of thicknesses of the layers.

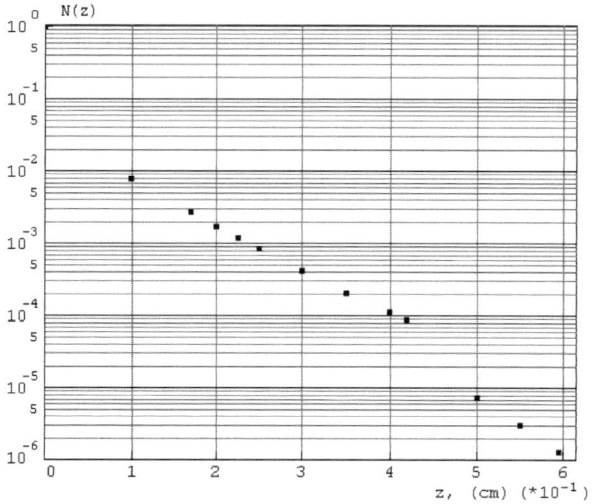

Figure 1. The amount of electrons as a function of the shield thickness for a heterogeneous $B_4Na_2O_7$ shield. Normalization to one particle. The Y-axis scale is logarithmic.

Numerical estimates of the amount of penetrated electrons (normalized to one particle) and mass attenuation coefficients for each layer are given in Table 3. Also given are the calculated characteristics of homogeneous shields. The mass attenuation coefficients appear to be close to each other, therefore our choice of the structure will be based on technological considerations. Boron-containing and nickel coatings can be deposited onto a finished housing of spacecraft equipment, which makes the heterogeneous shield more universal. In terms of the shield characteristics, all the discussed compounds are advantageous over aluminum.

Table 3. Layer-by-layer comparison of material properties in heterogeneous shields

Shield	Material	Density, kg/m^3	Thickness, m	Mass thickness, kg/m^2	Amount of penetrated electrons	Mass attenuation coefficient, m^2/kg
1	$B_4Na_2O_7$	1904	2.50E-03	4.76	6.62E-02	1.54
	Glass fabric	1596	1.70E-03	2.71	1.09E-01	0.82
	Ni	3898	1.75E-03	6.82	1.82E-02	1.66
	Entire shield	2156.4*	5.95E-03	12.83*	1.31E-06	1.06*
2	B_4C	1881	2.75E-03	5.17	5.87E-04	1.44
	Glass fabric	1596	1.70E-03	2.71	1.07E-01	0.82
	Ni	3898	1.75E-03	6.82	1.68E-06	0.60
	Entire shield	2134.4*	6.20E-03	13.23*	1.06E-06	1.04*

3	Homogeneous shield with $B_4Na_2O_7$	2156.4	5.95E-03	12.83	2.90-06	0.99
4	Homogeneous shield with B_4C	2134.4	6.20E-03	13.23	1.02E-06	1.04
5	Aluminum	2700	5.29E-03	14.29	2.641E-06	0.899

*– the mean value

When studying parameters of the protective shields under equally attenuated electron flow, we also calculated the secondary emission ratio (the amount of gamma-ray photons penetrated through all the layers) and the absorbed dose ratio. A 10^6 times attenuation of the electron flow is achieved with a 6 mm-thick aluminum shield. The absorbed dose for this shield thickness is 1.55E-11 μGy while this parameter for heterogeneous shields is an order of magnitude less. Table 4 gives the ratio of respective parameters between heterogeneous and aluminum shields.

We also analyzed the situation with homogeneous shields. A sodium tatraborate shield exhibits a 10% lower attenuation of the absorbed dose than aluminum, whereas heterogeneous shields are twice as efficient.

Table 4. Comparison of secondary emission and absorbed dose

Material	Thickness, mm	Amount of penetrated gamma-ray photons	Absorbed dose, μGy	Secondary emission ratio	Absorbed dose ratio
Aluminum	6	1.46E-3	1.55E-11	1	1
$B_4Na_2O_7$+T10+Ni	5.95	2.26E-4	7.89E-12	6.46	1.97
B_4C+T10+Ni	6.2	1.13E-4	6.60E-12	12.89	2.35
Homogeneous shield with $B_4Na_2O_7$	5.95	3.15E-4	1.65E-11	4.63	0.94
Homogeneous shield with B_4C	6.2	2.49E-4	6.06E-12	5.85	2.56

The shields under consideration let 4 to 5 times less gamma-ray photons resulting from Bremsstrahlung penetrate the shield, which is due to the heavy metal – nickel present in their composition. In Fig. 2 one can see that the amount of secondary emission particles drastically decreases in the interval from 4.2 mm, which corresponds to the nickel layer.

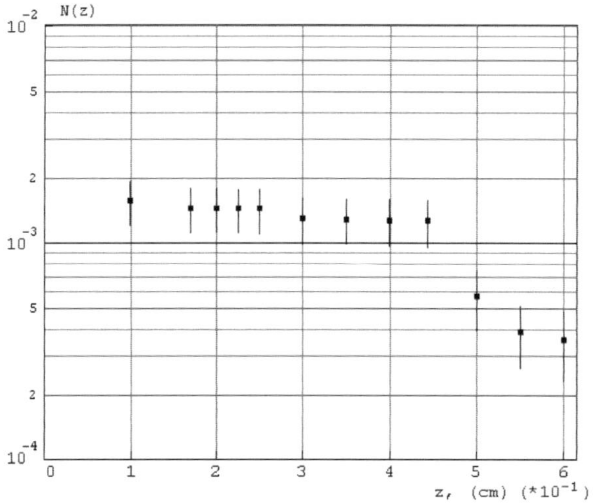

Figure 2. The amount of gamma-ray photons depending on the thickness of the B_4C heterogeneous barrier. Normalization to one particle. The Y-axis scale is logarithmic.

To compare the shields in terms of the amount of penetrated electrons, we calculated the real thickness of the aluminum layer (Table 5). It equals the mass thickness of a multilayered shield. A 5.3 mm-thick aluminum shield is 2 times less effective in trapping electrons than the heterogeneous shields under study.

Table 5. Finding the ratio of penetrated electrons

Material	Thickness, m	Amount of penetrated electrons	Mass attenuation coefficient, m^2/kg	Protection ratio
Aluminum	5.29E-03	2.64E-06	0.90	2.02
$B_4Na_2O_7$+T10+Ni	5.95E-03	1.30E-06	1.06	
B_4C+T10+Ni	6.2E-030	1.06E-06	1.04	1.97
Aluminum	5.45E-03	2.10E-06	0.89	

The mass of a protective shield is an important consideration as it affects the cost of satellite launching to orbit and as a consequence the market demand for this material. To make our estimates more specific, we adopted a model in the form of a 5 mm-radius cylinder. The cylinder height was as shown in Table 5. Comparison was carried out for the shields with the same amount of penetrated electrons (10^{-6} particles). The mass ratio between the heterogeneous and aluminum shields was 0.9 (Table 6). That is, on top of all the advantages discussed above, the shields under study are lighter than aluminum shields.

Table 6. Comparison of masses for the same amount of missed electrons, $N=10^{-6}$

Material	Mass,g	Mass ratio
Aluminum	127.23	1
$B_4Na_2O_7$+T10+Ni	112.27	0.88
B_4C+T10+Ni	115.51	0.91

We have considered and calculated using the "Computer laboratory" program [15] five different combinations of protective shields made of materials with high shielding quality and mass attenuation coefficient. Homogeneous and heterogeneous shields are advantageous over aluminum in lower bremsstrahlung and smaller weight. The boron carbide shields outmatch aluminum shields in all studied characteristics.

References.

1. Anashin V.S. [Problems of ensuring long active lifetime of REA of communication satellites] *Elektrosvyaz'.* 2009, No 4, P.19-22 (In Russ.)
2. Kuznetsov, N.V. *Radiatsionnaya opasnost' na okolozemnykh orbitakh i mezhplanetnykh traektoriyakh kosmicheskikh apparatov* [Radiation hazard on Earth orbits and interplanetary trajectories of spacecrafts] Available at: http://nuclphys.sinp.msu.ru/crd/crd3.htm. (In Russ.)
3. Novikov L.S. *Model' kosmosa. – 8-e izd. – T. 2: Vozdeystvie kosmicheskoy sredy na materialy i oborudovanie kosmicheskikh apparatov.* [Space model. – 8th prod. – T. 2: Impact of the space environment on materials and equipment of spacecrafts] *M.: Knizhnyy dom Universitet*, 2007, 1144 p. (In Russ.)
4. Pavlenko V. I., Edamenko O. D., Yastrebinskiy R. N. [Radiation protective composite material on the basis of a polystyrene matrix] *Vestnik BGTU im. V. G. Shukhova,* 2011, No 3, P. 113-116. (In Russ.)
5. Pavlenko V.I., Bondarenko G.G., Tarasov D.G., Edamenko O.D. [Modeling of impacts of fast electrons on polymeric radiation protective composites] *22 mezhdunarodnaya konferentsiya «Radiatsionnaya fizika tverdogo tela» pod red. Bondarenko G.G.* [Proceedings of the 22nd international conference "Radiation Physics of a Solid Body", Ed. by Bondarenko G.G.], Sevastopol' ,2012, P.222–228. (In Russ.)
6. Halbleib, J.A., Kensek, R.P., Valdez, G., Seltzer, S.M., Berger, M.J. ITS: the integrated TIGER series of electron/photon Monte Carlo transport codes – Version 3.0. *IEEE Trans. Nucl. Sci.* , 1992, Vol. 39 (4), P. 1025–1029.
7. Tatsuo Tabata, Vadim Moskvin, Pedro Andreo, Valentin Lazurik, Yuri Rogov. Extrapolated ranges of electrons determined from transmission and projected-range straggling curves. *Radiation Physics and Chemistry,*2002, No 64, P.161–167.
8. J. Bell, C. Martin, D. Lail, P. Nguyen. Radiation shielding for a lunar base. *National Aeronauticsand Space Adiminstratoin*, 2011.
9. Bezrodnykh I. P., Morozova E. I., Petrukovich A. A. *Tormoznoe izluchenie elektronov v veshchestve kosmicheskogo apparata. Metodika rascheta, T.120.* [Brake radiation of electrons in the spacecraft material. Calculation procedure.] Voprosy elektromekhaniki, 2011, P. 37–44. (In Russ.)
10. Novikov L.S., Mileev V.N., Voronina E.N., Galanina L.I., Makletsov A.A., Sinolits V.V. [Radiative effects on materials of space equipment. Surface. X-ray, synchrotron and neutron studies.] *Rentgenovskie, sinkhrotronnye i neytronnye issledovaniya*, 2009, No. 3, P. 1– 18. (In Russ.)

11. Bespalov V.I. *Lektsii po radiatsionnoy zashchite : ucheb.posobie*. [Lectures on radiation protection: textbook.] Tomsk : Izd-vo Tomskogo politekhnicheskogo universiteta, 2012. P. 40-42. (In Russ.)
12. Kimel' L.R., Mashkovich V.P. *Zashchita ot ioniziruyushchikh izlucheniy. Spravochnik. Izd.2.*[Protection against ionizing radiation. Reference book. Edition 2.] M.: Atomizdat, 1972, 312 p. (In Russ.)
13. Khaffner Dzh. *Yadernoe izluchenie i zashchita v kosmose.* [Nuclear radiation and protection in space] M. :Atomizdat, 1979. 304 p. (In Russ.)
14. Varga L., Horvath E, Evaluation of electronics shielding in micro-satellites. Technical memorandum: DRDC Ottawa TM 2003-017. Defence R&D Canada. *Ottawa*, 2003. 25 p.
15. Bespalov V.I. [An EPHCA software package for statistical modeling of radiation fields of photons and charged parts.] *Izv. Vuzov Fizika, Prilozhenie,* 2000, No. 4, P. 159–165. (In Russ.)

Combined string searching algorithm based on knuth-morris-pratt and boyer-moore algorithms

R Yu Tsarev[1], A S Chernigovskiy[1], E A Tsareva[2], V V Brezitskaya[2], A Yu Nikiforov[1], N A Smirnov[2]
[1]Siberian Federal University, Krasnoyarsk, Russia
79, Svobodny Prospect, Krasnoyarsk, Russia
[2]Siberian State Aerospace University, Krasnoyarsk, Russia
31 "Krasnoyarskiy Rabochiy" prospect, Krasnoyarsk, 660037, Russia

E-mail: tsarev.sfu@mail.ru

Abstract. The string searching task can be classified as a classic information processing task. Users either encounter the solution of this task while working with text processors or browsers, employing standard built-in tools, or this task is solved unseen by the users, while they are working with various computer programmes. Nowadays there are many algorithms for solving the string searching problem. The main criterion of these algorithms' effectiveness is searching speed. The larger the shift of the pattern relative to the string in case of pattern and string characters' mismatch is, the higher is the algorithm running speed. This article offers a combined algorithm, which has been developed on the basis of well-known Knuth-Morris-Pratt and Boyer-Moore string searching algorithms. These algorithms are based on two different basic principles of pattern matching. Knuth-Morris-Pratt algorithm is based upon forward pattern matching and Boyer-Moore is based upon backward pattern matching. Having united these two algorithms, the combined algorithm allows acquiring the larger shift in case of pattern and string characters' mismatch. The article provides an example, which illustrates the results of Boyer-Moore and Knuth-Morris-Pratt algorithms and combined algorithm's work and shows advantage of the latter in solving string searching problem.

1. Introduction

The searching problem is one of fundamental tasks of theoretical programming [10]. String searching is one of simple, but, nonetheless, extremely important problems. The importance of this problem is explained by the wide area of its solution results' application: text editors, online string matching, speech analysis and recognition, information retrieval, network content inspection, data compression, etc. [4].

Nowadays there are many algorithms for solving this problem developed. During the last thirty years numerous algorithms, which allow solving the problems of searching with some special characteristics, were offered. M. Ahmed, M. Kaykobad and R.A. Chowdhury developed a new string matching algorithm that unlike other sub-linear string-matching algorithms never performs more than n text character comparisons while working on a text of length n [1]. T. Lecroq proposed string matching algorithms based on hashing q-grams [9]. K. Fredriksson and S. Grabowski developed new exact bit-parallel string matching algorithm, based on the shift-or algorithm [2, 5]. They have shown how to adapt their techniques for the shift-add algorithm, obtaining optimal time for searching under Hamming distance [5]. L. He, B. Fang and J. Sui have used the concept of a window whose size is equal to pattern length. They presented a novel string matching algorithm named wide window algorithm [6]. A.

Hudaib et al. in [7] presented two sliding windows algorithm. The algorithm makes use of two sliding windows. Both windows slide in parallel over the text until the first occurrence of the pattern is found or until both windows reach the middle of the text.

The inexhaustible interest towards this sphere of problems confirms the importance and topicality of the string searching problem. This article offers an algorithm based upon two well-known and acknowledged string searching algorithms. This kind of combination allows increasing the effectiveness of solving the given problem.

2. Theoretical bases of the combined algorithm

The combined string searching algorithm is based upon two algorithms. One of them was developed by Knuth, Morris and Pratt, and the other one by Boyer and Moore. These algorithms belong to two quite big sub-classes of string searching algorithms. Knuth-Morris-Pratt algorithm belongs to forward pattern matching sub-class, and Boyer-Moore belongs to backward pattern matching sub-class. Two stages can be identified in the work of these algorithms:

1. Preparing of table d used in case of shifting of the pattern in the string.
2. String searching itself.

Let us have a look at the work of these string searching algorithms in detail. The description will be followed by examples and explanations with the help of C language.

2.1. Principles of Knuth-Morris-Pratt string searching algorithm

The algorithm was developed by D. Knuth and V. Pratt, and independently by J. H. Morris in 1974, but it was published by them collaboratively only in 1977 [8]. This algorithm is based upon the idea that after partial matching of the initial part of the pattern with the corresponding characters of the string, we get certain information on the basis of the pattern itself, and this information will allow us to move forward along the string not by one, as in case of naive string matching, but further. In order to do that, when shifting, Knuth-Morris-Pratt string searching algorithm (or KMP algorithm) uses table d, which is preprocessed before the beginning of string searching. Since table d preprocessed in compliance with KMP algorithm will also be used in combined algorithm, let us denote it as d_{KMP}.

Thus, table d_{KMP} for Knuth-Morris-Pratt string searching algorithm is preprocessed on basis of the pattern and contains values, which will be used further for calculation of the shift amount of the pattern. The size of the table is equal to the length of the pattern. Therefore, table d_{KMP} is in fact a one-dimensional array, consisting of the number of elements equal to the number of characters in the pattern.

The first element of the array d_{KMP} is always minus one.

Let us have a look at the formation of the table d_{KMP} in the example of the pattern “barbarian”. As it is shown in Figure 1, the first element of the table d, corresponding to the first character of the pattern “b” is equal to minus one.

b	a	r	b	a	r	i	a	n
–1								

Figure 1. Pattern and value of the first element of table d_{KMP}.

For the other characters, the values of the elements of table d_{KMP} are calculated in the following way. The value of $d_{KMP}[j]$ corresponding to character j of the pattern is equal to the maximum number of characters, which directly precede the given character and match with the beginning of the pattern. If k characters precede the given character, then only k−1 preceding characters are taken into account.

Thus, Figure 2 shows that one character "b" directly precedes the fifth character "a", and matches with character b at the beginning of the pattern. Both characters "b" are in bold type. Since the number of matching characters is equal to one, element of the table, corresponding to the fifth symbol of the pattern "a" receives value one (see Figure 2).

b	a	r	**b**	a	r	i	a	n
–1				1				

Figure 2. Pattern and value of the fifth element of table d_{KMP}.

Two characters "ba" directly precede the sixth character of the pattern r and match with the characters "ba" at the beginning of the pattern. Both pairs of characters "ba" are in bold type. Since the character "r" is preceded by two characters, matching with the first two characters of the pattern, the corresponding element of the table is given the value 2 (see Figure 3).

b	**a**	r	**b**	**a**	r	i	a	n
–1				1	2			

Figure 3. Pattern and value of the sixth element of table d_{KMP}.

No other characters of the pattern "barbarian" are preceded by characters, matching with the beginning of the pattern; therefore the corresponding elements of the table d_{KMP} are equal to zero (see Figure 4).

b	a	r	b	a	r	i	a	n
–1	0	0	0	1	2	0	0	0

Figure 4. Pattern and table d_{KMP}.

The second stage of KMP algorithm work is comparing characters of the string and characters of the pattern and calculating the shift amount in case of their mismatch. The characters are considered from the left to the right, i.e. from the beginning to the end of the pattern. In case of mismatch of characters of pattern and string, the pattern is shifted to the right along the string by following value:

$$j - d_{KMP}[j],$$

in which j is an index of the currently considered character of the pattern, $d_{KMP}[j]$ is the value of the table d_{KMP}, corresponding to this character.

Figure 5 shows an example illustrating the shift of the pattern during the work of KMP algorithm.

String	*B*	*a*	*r*	*b*	*e*	*r*	...		
Pattern	*B*	*a*	*r*	*b*	*i*				
j	0	1	2	3	4				
$d_{KMP}[j]$	−1	0	0	0	0				
					B	*a*	*r*	*b*	*i*

$j = 4$, $d_{KMP}[4] = 0$,
shift $= j - d_{KMP}[j] = 5$

Figure 5. Partial match and shift of the pattern.

Let us give an example of string searching of the pattern "barbarian". Figure 6 shows the process of KMP algorithm work. The characters that are being compared are underlined.

```
b a r   i s   f u l l   o f   b a r b a r i a n s
b a r b a r i a n
      b a r b a r i a n
        b a r b a r i a n
          ...
                            b a r b a r i a n
                              b a r b a r i a n
```

Figure 6. Process of work of Knuth-Morris-Pratt string searching algorithm.

Pay attention to the fact that in every case of characters' mismatch, the pattern is shifted by the whole passed length, because smaller shifts cannot lead to full match.

As far as the effectiveness of Knuth-Morris-Pratt string searching algorithm is concerned, its developers show that about n + m character comparisons are needed, which is better that n * m comparisons in case of naive string matching (where n is length of the string, m is length of the pattern, $0 < m < n$) [8, 10]. In this case, the string scanning indicator i never goes back, while in case of naive string matching, after a mismatch, the considering starts again from the first character, and therefore the string characters that have been considered before might be processed again.

However, string searching with the use of Knuth-Morris-Pratt string searching algorithm is helpful only in case if the mismatch of string and pattern characters was preceded by some number of matches. If the comparison of the string with the pattern shows that first characters are different, the pattern shifts by only one character. The pattern shifts more than by one only in case if several characters of string and pattern match.

2.2. Principles of Boyer-Moore string searching algorithm

In 1977 R.S. Boyer and J.S. Moore offered an algorithm, which not only improves the processing of the worst case, but also gives and advantage in general case [3]. Boyer-Moore string searching algorithm (or BM algorithm) is based upon an unusual idea –characters' comparing starts not from the beginning, but from the end of the pattern. The speed of Boyer-Moore algorithm's running is achieved at the expense of omitting parts of the text, which certainly do not participate in successful comparison.

As well as in case of KMP algorithm, table d is preprocessed basing upon the pattern before the beginning of searching, and then used when shifting of the pattern along the string.

Since this table will also be used in the work of the combined algorithm, let us denote it as d_{BM}.

Initially all the elements of the table d_{BM} are given the values equal to the length of the pattern.

On the next stage, every element of the table d_{BM} is given a value equal to the remoteness of the corresponding character of the pattern from the end of the pattern. Figure 7 shows the pattern and remoteness of each of its characters from the end of the pattern.

b	a	r	b	a	r	i	a	n
8	7	6	5	4	3	2	1	

Figure 7. Pattern and remoteness of each of its characters from the end of the pattern.

In case if a pattern contains several identical characters, the element of table d_{BM} corresponding to this character is given the value equal to remoteness from the end of the pattern of the rightmost character [10]. For example, the pattern "barbarian" contains three characters "a", and remoteness of the rightmost from the end of the pattern is equal to one. In connection with this, the elements of table d_{BM} corresponding to the rest of characters "a" in the pattern are given the value equal to one (see Figure 8). Similarly, elements of table d_{BM} corresponding to all the characters "r" are given the value of three, and the elements corresponding to characters "b" will have the value of five.

b	**a**	r	b	**a**	r	i	**a**	n
5	**1**	3	5	**1**	3	2	**1**	

Figure 8. Pattern and table d_{BM}.

During the implementation of Boyer-Moore string searching algorithm, as well as during the implementation of the combined algorithm, it is recommended to use the table of character codes ASCII for creation of table d_{BM}.

Every printed character has its own ASCII code. For example, the code of the character "a" is 97, the code of "r" is 114, and the code of space is 32. It can also be mentioned that the codes of Cyrillic characters are situated between 192 and 255.

Table ASCII contains 256 characters, that is why table d_{BM} can be named a one-dimensional integer-valued array consisting of 256 elements: int d[256].

Let us have a look at the peculiarities of software implementation of table d_{BM} through the example of the pattern "barbarian". Since this pattern contains nine characters, its length is nine. Correspondingly, on the stage of initialization all the elements of the table d_{BM} are given the value of nine:

$$d[0] = d[1] = \ldots = d[254] = d[255] = 9.$$

Then elements of table d_{BM} are given the corresponding values equal to the remoteness of the given character from the end of the pattern. The index of the element of table d_{BM} that is given a new value is identified by the ASCII code of the considered character.

For example, ASCII code of character "a" is 97. The value of the elements of table d_{BM} corresponding to it, as it was already shown on Figure 9, is equal to one. Therefore:

d[97] = 1.

It can also be written in the C language:

d['a'] = 1.

It can also be mentioned that during the implementation the element of array d with the index of 97 corresponds to all the characters "a" in the pattern "barbarian". Thus, all the elements of table d_{BM} corresponding to the characters "a" of the pattern in such case are given the value of 1 (see Figure 9).

The values of the elements of table d_{BM} corresponding to characters "i", "r", "b" change in the similar way:

d['i'] = 2 or d[105] = 2,
d['r'] = 3 or d[114] = 3,
d['b'] = 5 or d[98] = 5.

Finally, the element of table d_{BM} corresponding to the character situated in the end of the pattern should be given the value of one:

d['n'] = 1 or d[110] = 1.

The necessity of assigning value of one and not zero (since its remoteness from the end of the pattern is equal to zero) is connected with the peculiarities of calculation of shift amount directly during the string searching.

The second stage of BM algorithm work is string searching itself. While comparing the pattern and the string, the pattern moves from the left to the right along the string. However, pattern and string characters are compared from the right to the left along the pattern. That is the peculiarly of backward pattern matching.

The comparison of the pattern and the string is carried out 1) until the whole pattern is considered, which indicates that there is a match between the pattern and some part of the string, 2) until the string ends, which means that there are no entries matching the pattern in the string, 3) until there is a mismatch of the pattern and string characters, which leads to shifting of pattern by several characters to the right and continuing the searching process.

In case of characters' mismatch, shift of the pattern along the string is defined by the value of the element of table d_{BM}. However, the index of the given element is the ASCII code of the character of the string. It can be emphasized that although array d is formed on the basis of the pattern, the shift is defined by the mismatching character of the string. Figure 9 shows an example of BM algorithm work. The characters that are being compared are underlined.

```
b a r   i s   f u l l   o f   b a r b a r i a n s
b a r b a r i a n
                  b a r b a r i a n
                        b a r b a r i a n
                              b a r b a r i a n
```

Figure 9. Process of work of Boyer-Moore string searching algorithm.

In the first iteration there was a mismatch of the pattern character n and string character u. It should be pointed out that the values of all the elements of array d for all the possible characters on the stage of initialization are equal to the length of the pattern, i.e. to nine. Since character "u" is not found in the pattern, the value of the element d['u'] in further formation of the table is not changed and remains equal to nine. That is why the pattern is shifted by nine characters to the right. If these two characters matched, the last but one pattern character and the corresponding string character would be considered next, etc.

While comparing the pattern and the string there is a mismatch of characters "n" and "r". Once again, defining the shift with the help of the string character (d['r']), we get the value of three. The pattern shifts to the right by three characters. Similarly, the pattern gradually shifts along the string, until the pattern is found in the string or until the string ends.

The evaluation of BM algorithm effectiveness shows that almost in all cases the algorithm demands far less n comparisons. In the most favorable conditions, when the last pattern character falls into the mismatching string character, the number of comparisons is n / m [3, 10].

2.3. *Combination of Knuth-Morris-Pratt and Boyer-Moore string searching algorithms*

Based upon the mentioned principles of Knuth-Morris-Pratt и Boyer-Moore string searching algorithms concerning the creation of tables for defining the shift and string searching itself the following combined algorithm has been offered.

Step 1. Creating table d_{KMP} according to the principles of Knuth-Morris-Pratt string searching algorithm.

Step 2. Creating table d_{BM} according to the principles of Boyer-Moore string searching algorithm.

Step 3. Defining the initial value of index i, corresponding to the position of the pattern relative to the string.

Step 4. Defining the initial values of indices j_{KMP} and j_{BM}, which indicate the beginning and the end of the pattern respectively.

Step 5. Comparing the pattern character with index j_{KMP} and the corresponding string character, comparing the pattern character j_{BM} and the corresponding string character. If at least one comparison ends with a mismatch, go to Step 11.

Step 6. If j_{KMP} is less than j_{BM}, go to Step 9.

Step 7. Output of the message informing that the pattern matches with a part of the string.

Step 8. Go to Step 15.

Step 9. Increasing index j_{KMP} by one (i.e. going to the next character to the right), decreasing index j_{BM} by one (i.e. going to the next character to the left).

Step 10. Go to Step 5.

Step 11. Choosing the larger shift from $j_{KMP} - d_{KMP}[j_{KMP}]$ and $d_{BM}[i + \text{pattern length} - j_{BM}]$.

Step 12. Shifting the pattern to the right relative to the string, increasing the value of index i by the shift defined in Step 11.

Step 13. If the sum of index i and the length of the pattern is less than the length of the string, go to Step 4.

Step 14. Output of the message informing that the target pattern is not found.

Step 15. Stop.

The combined algorithm allows finding matches between the pattern and a part of the string with a smaller number of shifts, due to the fact that in Step 11 a larger shift is chosen out of two shifts, which can be received with Knuth-Morris-Pratt and Boyer-Moore string searching algorithms. At the same time, these algorithms themselves guarantee that however large the shift is, no match of the pattern and the string will be overlooked.

3. Results and discussion

The combined string searching algorithm based upon Boyer-Moore and Knuth-Morris-Pratt string searching algorithms was implemented within computer programme InfoSearch. This programme allows carrying out the analysis of work of Boyer-Moore, Knuth-Morris-Pratt algorithms and the combined algorithm. The interface of the programme is shown in Figure 10.

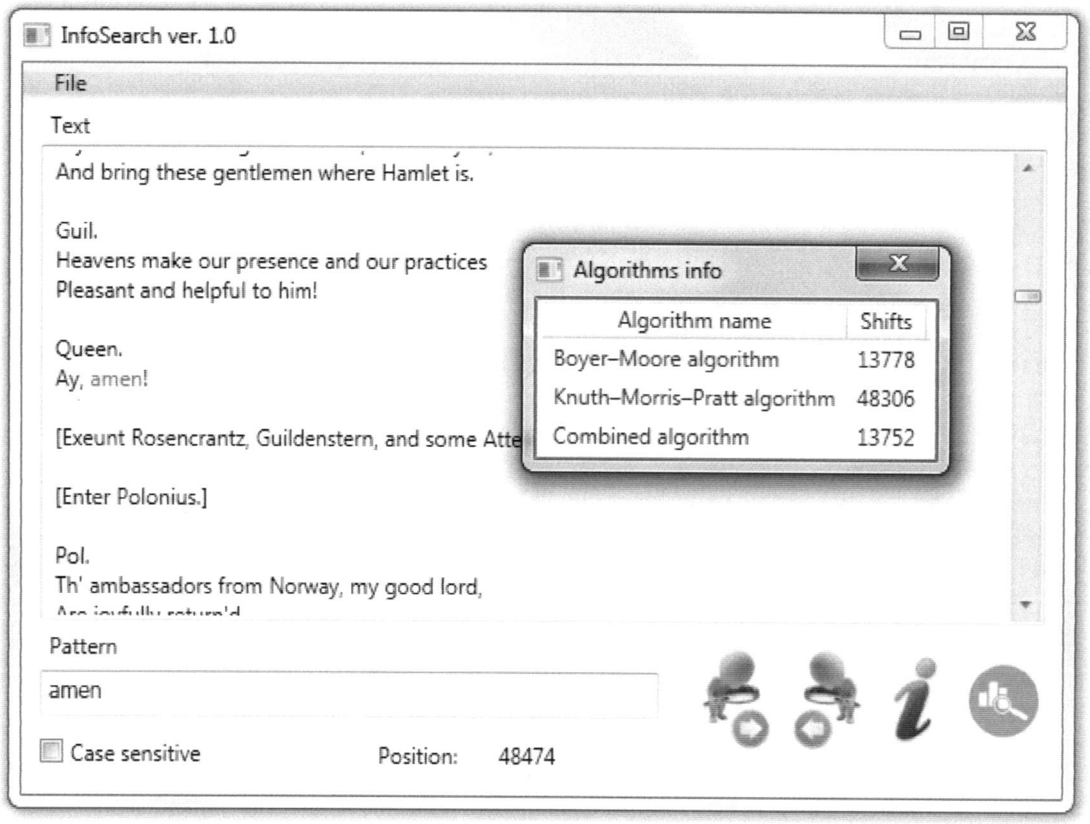

Figure 10. Interface of the programme implementing the combined string searching algorithm.

The analysis of the considered string searching algorithms work results can be shown in the example of The Tragedy of Hamlet, Prince of Denmark written by William Shakespeare. Table 1 shows the results of the algorithms when searching for the words written in the first column.

Table 1. Results of string searching algorithms

Word	Position	Shifts		
		Knuth-Morris-Pratt algorithm	Boyer-Moore algorithm	Combined algorithm
amen	48474	48306	13778	13752
antique (1st occurrence)	68539	67756	11676	11667
antique (2nd occurrence)	179186	177135	30677	30646
cozenage	166058	165546	24932	24921
habit (1st occurrence)	24743	24519	5738	5729
habit (2nd occurrence)	29215	28949	6762	6751
habit (3rd occurrence)	113830	112669	26005	25973
habit (4th occurrence)	171586	169831	39245	39204
herb	136155	133534	37481	37423
marble	30208	30025	5985	5948
marvel	18318	18193	3624	3592
matron	111378	110925	22596	22588
theme (1st occurrence)	14063	13587	3171	3161
theme (2nd occurrence)	161372	156067	36456	36390
thieves	138812	135466	23527	23480
sea-fight	165436	164562	23227	23177
stone (1st occurrence)	129310	128472	31382	31341
stone (2nd occurrence)	140499	139599	34100	34057

The analysis that was carried out reveals that the combined algorithm needs by several times less shifts than Knuth-Morris-Pratt algorithm to find an entry of the pattern into the string.

The combined algorithm also shows better results in terms of required number of shifts in comparison with Boyer-Moore algorithm, although this time the advantage is not as vivid as the advantage of the combined algorithm over Knuth-Morris-Pratt algorithm.

4. Conclusion

This article presents a solution of a task, which is extremely important for computer analysis – namely, string searching. The offered solution develops theoretical basis of methods for language data computer analysis and serves the purposes of solving practical computer linguistics tasks.

The developed combined string searching algorithm has united the advantages of well-known Knuth-Morris-Pratt and Boyer-Moore string searching algorithms. The distinction of combined algorithm lies in higher effectiveness in comparison with the initial algorithms and larger shifts in case of mismatch of string and pattern characters, which increases the speed of pattern entry in the string. The combined string searching algorithm can be successfully implemented for searching in English and Russian texts.

5. Acknowledgements

The authors would like to thank Ahmad Hassanat from Mu'tah University, Jordan, for consultations while studying string searching algorithms.

References

[1] Ahmed M, Kaykobad M and Chowdhury R A 2003 A new string matching algorithm *Int. J.*

Comput. Math. **80** 825–834

[2] Baeza-Yates R A and Gonnet G H 1992 A new approach to text searching *Commun. ACM* **35** 74–82

[3] Boyer R S and Moore J S 1977 A fast string searching algorithm *Commun. ACM* **20** 762–772

[4] Faro S and Lecroq T 2013 The exact online string matching problem: A review of the most recent results *ACM Comput. Surv.* **45**

[5] Fredriksson K and Grabowski S 2005 Practical and optimal string matching *Lecture Notes in Computer Science (including subseries Lecture Notes in Artificial Intelligence and Lecture Notes in Bioinformatics)* **3772 LNCS** 376-387

[6] He L, Fang B and Sui J 2005 The wide window string matching algorithm *Theor. Comput. Sci.* **332** 391-404

[7] Hudaib A, Al-Khalid R, Suleiman D, Itriq M and Al-Anani A 2008 A fast pattern matching algorithm with two sliding windows (TSW) *J. Comput. Sci.* **4** 393-401

[8] Knuth D, Morris J H, Pratt V 1977 Fast pattern matching in strings *SIAM J. Comput.* **6** 323–350

[9] Lecroq T 2007 Fast exact string matching algorithms *IPL* **102** 229-235

[10] Wirth N 1985 *Algorithms and data structures* (NJ: Prentice Hall)

Using GLONASS for precise determination of navigation parameters under interference from various sources*

V N Tyapkin[1] **Yu L Fateev**[2] **D D Dmitriev**[3]**I N Kartsan**[4] **P V Zelenkov**[5] **A E Goncharov**[6] **I R Nasyrov**[7]

[1] Associate professor, Siberian Federal University, Krasnoyarsk, Russia
[2] Professor, Siberian Federal University, Krasnoyarsk, Russia
[3] Associate professor, Siberian Federal University, Krasnoyarsk, Russia
[4] Associate professor, Reshetnev Siberian State Aerospace University, Krasnoyarsk, Russia
[5] Science and Research Department Director, Reshetnev Siberian State Aerospace University, Krasnoyarsk, Russia
[6] Associate professor, Reshetnev Siberian State Aerospace University, Krasnoyarsk, Russia
[7] Post-graduate student, Reshetnev Siberian State Aerospace University, Krasnoyarsk, Russia

E-mail: kartsan2003@mail.ru

Abstract. This article discusses the main approaches to the designs of systems for determining location and spatial attitude based on satellite navigation equipment. The article describes possible solutions for constructing an angular attitude measurement system capable of spatial interference selection on the basis of a single antenna system.

Keywords: object angular attitude, phased antenna array, interference stability, GLONASS, algorithms and equipment of broadband interference suppression

1. Introduction

Modern signal receivers utilized in global satellite radionavigation systems have one significant drawback – low immunity to interference. The modernization of existing receivers by reducing the bandwidths of their tracking systems, integral tracking of a full set of signals from navigation satellites at all frequency bands, narrowband filtering of (in the limit – harmonic) interferences, providing direct entry into the synchronization with a high precision GLONASS signal (this is effective against narrowband interference), and other methods enable the operation of navigation equipment with an interference to signal relation at input of the receiving antenna at no more than 60 – 65 dB. Further improvements to interference immunity for navigation equipment (to a level of 80 – 85 dB) are possible only with the application of spatial methods to resist interference.

Nowadays the widespread implementation of satellite navigation systems requires substantial improvement of equipment responsible for interference immunity. This includes deliberate interference (jamming). Estimates show that the required level for interference immunity to broadband interference in the useful signal band is $J/S \approx 90\ dB$. Although there

* This research has been financed by the Ministry of Education and Science of the Russian Federation at Siberian Federal University (Contract № 14.578.21.0021).

are many methods for suppressing interference, they are all based on utilizing distinctions (amplitude, temporal, frequency, angular, polarizing, etc) in useful and disturbing signals. The idea of making these distinctions useful consists in the formation of minima (nulls) in angular, polarizing, frequency and other receiver parameters in corresponding directions towards the source of interference and maxima in direction of a useful signal. This suppresses or compensates the interference and accumulates the useful signal.

The final effect of the processing is defined by the level and completeness of use of available distinctions and the corresponding methods, their implementation into the unit diagram models for the satellite navigational user equipment.

The following approaches should be considered as the main methods for improving interference immunity for the studies equipment [1-7]:

- reducing the tracking systems bandpass; these systems measure satellite signal parameters using available aprioristic data on the dynamic movements of the user antenna system (one source of such data can be inertial navigation devices which is found in the satellite navigation user equipment);
- integral tracking for a full set of signals from the satellite at all frequency bands;
- filtering narrow-band (at its limit – harmonic) interference in the signal to correlator processing chain by means of adaptive digital filters or by utilizing spectral filtration; this includes measuring the parameters of interference and its compensation in the digital sector of the equipment to the correlators;
- providing direct access into the synchronization with a high precision GLONASS signal (effective for narrow-band interference).

Estimates [1] show that the aforementioned methods may provide the operability of satellite navigation user equipment at the following level for interference / signal relation upon input of the receiving antenna – $J/S \approx 60 \ldots 65\ dB$. Further interference stability improvement to $J/S \approx 80 \ldots 90\ dB$ is possible by using spatial methods for interference suppression. These methods require the spatial structure of signal wave fronts, generated either by navigation satellites or by the source of interference, to be tracked as do methods for determining spatial attitude (orientation). Regarding this, the anti-interference and the angular receivers should be equipped with antenna arrays. Tracking the structure of spatial signals on the array elements enables us to reduce the multibeam reception error. Thus, in the long term, the angular receiver should be developed on the basis of a low-element antenna array (6 to 12 elements). This will provide a solution for the following triune problem:

1) determining spatial attitude of objects;
2) increasing interference stability amplification with spatial selection methods;
3) suppressing multibeam signal distribution.

The design for of such equipment may vary in terms of antenna array configuration and the receiving and measuring unit layout.

2. Application of three identical low-element antenna arrays dispersed in space.

The simplest layout for anti-interference equipment capable of detecting spatial attitude has been suggested by V. N. Kharisov [1]. It is based on three identical low-element antenna arrays (Fig. 1) that are dispersed in space. A design like this is the modernization of a typical angular antenna system where each antenna has been replaced with an anti-interference antenna array.

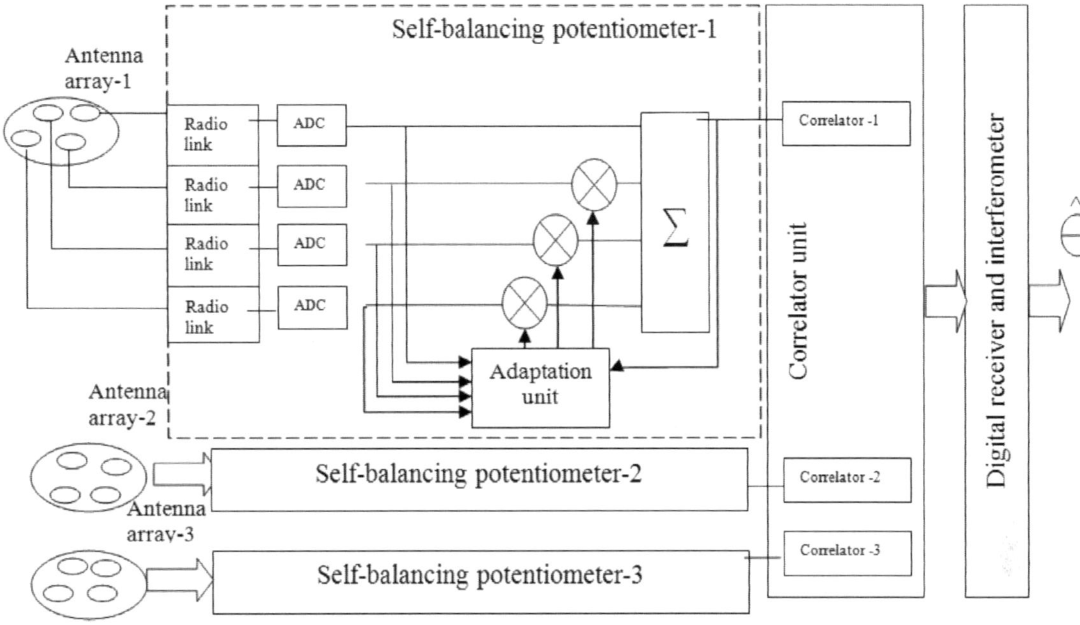

Figure 1. Anti-interference angular satellite navigation user equipment based on three identical low-element antenna arrays.

For interference suppression the same weighting coefficients are used in each antenna array as the interference situation is identical for all elements of the interferometer. In this case, the carrier phase shifts equally for all antenna arrays; the phase shifts between output signals are identical to single antennas.

3. Application of an angular system antenna array for interference suppression

The main drawback of this first anti-interference antenna system for navigation equipment, capable of determining spatial attitude, is the excessive number of antennas. The system is equipped with 12 antennas. However during interference suppression only 4 antennas are employed for each case; for spatial attitude determining there are only two bases – this is equivalent to the application of three antennas. If N number of antennas is required for determining spatial attitude and M number of antennas are required for interference suppression then the total number of antennas for this variant is $N \times M$, i.e. the low application efficiency of the antenna array is obvious.

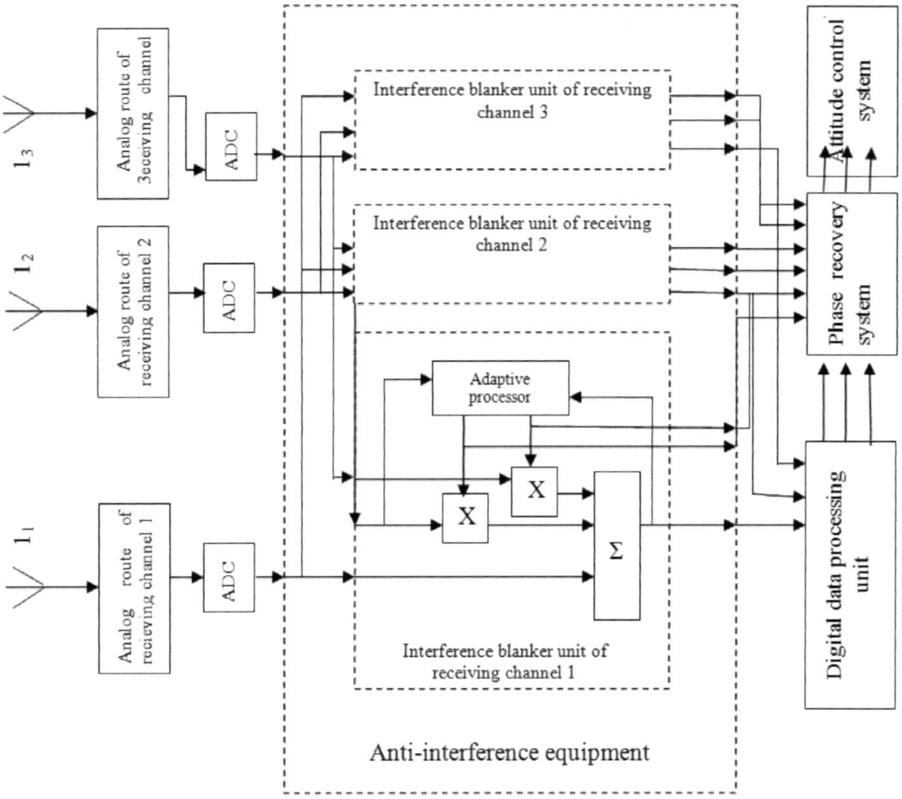

Figure 2. Layout of angular satellite navigation user equipment using interferometer antennas as additional channels for self-balancing potentiometers.

To increase the exploitation efficiency of the antenna array, another layout (Fig. 2) is suggested. It has several interference suppression routes equipped with an allocated channel, different reference antennas [5] are used for each channel. The remaining antennas of the array are used as additional channels for interference suppression.

4. Building an angular system on the basis of antenna arrays with a large number of elements

In the first two proposed layouts, a minimum number of antenna array elements were supposed to be utilized. However, nowadays we have an opportunity to build antenna arrays with large numbers of elements – up to 8 or 12. Due to the redundancy of the antenna system, this allows us to further develop methods for both interference suppression and determining spatial attitude.

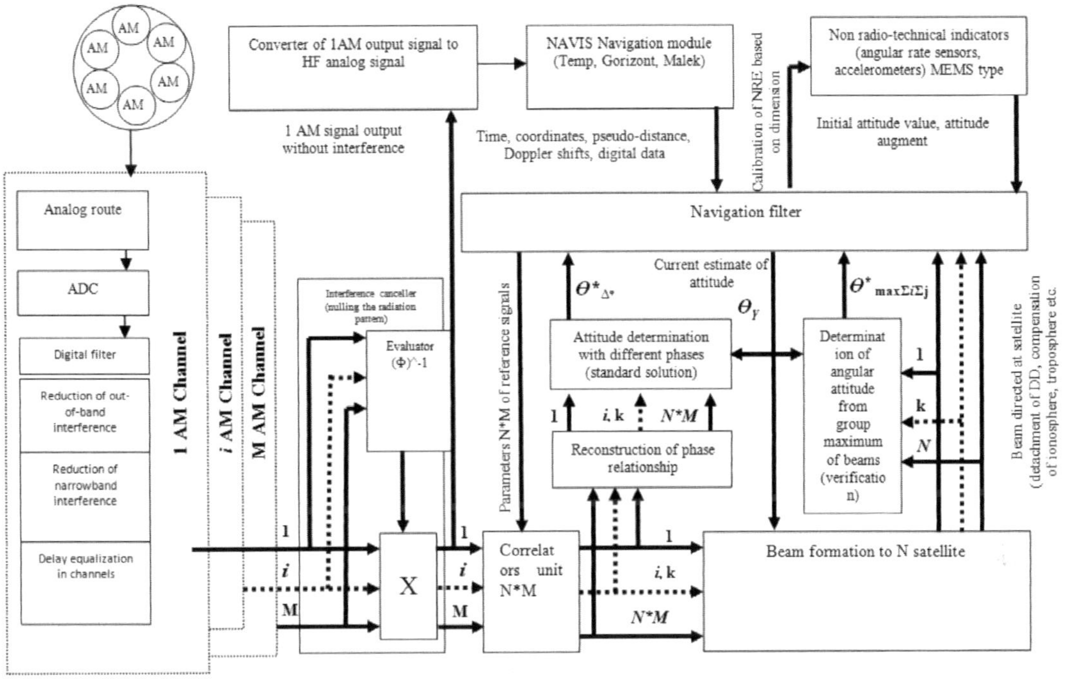

Figure 3. Layout of angular satellite navigation user equipment based on antenna arrays with a large number of elements.

The redundancy of the antenna array can be used for eliminating additional diffraction nulls during interference suppression and to improve accuracy and reliability when determining spatial orientation.

A drawback of the third variant is the equipment becoming increasingly overcomplicated for both antenna array and the receiving and measuring unit.

However, the application of arrays for a simultaneous solution of the aforementioned problem for all proposed layouts demands some features to be considered.

When the anti-interference radionavigation equipment is arranged, one antenna is always added to the minimum number of antennas: there is always one antenna more than the estimated number of suppressed interference signals. So, if it is necessary to suppress interference from three directions the antenna array should contain at least four antennas. The interference suppressor represents a complex gravimetric adder of signals from all elements of the array. Weighting coefficients are selected so that interference signals are completely compensated during addition.

The arrangement of satellite navigation equipment that is capable of determining spatial attitude requires no less than a triple antenna interferometer.

As the research has clearly demonstrated, the number of antennas should be no less than $n+2$ and $n+3$, in certain cases, for the design of anti-interference satellite navigation user equipment capable of determining spatial attitude.

To determine the angles, the phase shift of signals between antennas is measured. This determines the spatial attitude of an object.

The first two layouts require utilizing a null beamformer in the directions of the interference source. Another requirement is for the phase of satellites signals to be corrected after adapting. The third layout suggests the employment of a beamformer in direction of the satellites; in this case it is necessary to be aware of the spatial configuration of the antenna system.

5. Conclusion

Thus, a theoretical justification for the principal decisions on improving the accuracy and interference immunity of the developed modifications for angular receivers allows us to make the following conclusions:

Anti-interference modifications should be built on the basis of the adaptive antenna arrays that include the functions of a null beamformer pointed in the direction of the source of interference. They provide a considerable (exceeding 35 dB) reduction of interference; it is possible to achieve a higher amplification of the signal (≈8 ... 15 dB) if the beamformer function of the radiation pattern is realized in the direction of each satellite. This will also minimize the error from the multibeam reception signal, which proves to be an integral part of the error for determining spatial attitude.

References

[1] GLONASS Pritsipy postroeniia i funktsionirovaniia (Principals of arrangement and functioning) 2010 Radiotekhnika (Moscow, Russia) Petrov A I, Kharisov V N ed.

[2] Tyapkin V N, Dmitriev D D, Garin E N, Sokolovskii A V 2013 Controlling amplitude-phase distribution with an adaptive phase antenna array *Promyshlennye ASU i kontrollery* No. 4 P 66-71

[3] Fateev Yu L, Chmykh M K Methods of attitude control using signals from satellites in global satellite navigation systems Patent RUS 2105319

[4] Tyapkin V N 2012 Identifying angular attitude of object using radionagigation signals from satellites under deliberate broadband interference *Uspekhi sovremennoi radioelektroniki. Zarubezhnaia radioelektronika* No. 9 p 100-106

[5] Avsievich V N, Grebnikov A V, Kokorin V I 2000 Opyt sozdaniia i perspektivy primeneniia apparatury potrebitelei global'nykh sputnikovykh sistem (Experiences of developing and persectives of utilizing global navigation satellite user equipment systems) *Giroskopiia i navigatsiia No. 4 p 104*

[6] Markel M, Sutton E, Zmuda H 2001 An antenna array-based approach to attitude determination in a jammed environment *Proceedings of the ION GPS '01* (Salt Lake City, Utah)

[7] Ugazio S, Lo P, Falletti E 2013 Multipath Mitigation Using Linear Adaptive Filtering Techniques *Proceedings of the 26th International Technical Meeting of The Satellite Division of the Institute of Navigation (ION GNSS+ 2013)* (Nashville, Tennesse) p 2007-2014

IOP Conf. Series: Materials Science and Engineering **122** (2016) 012036 doi:10.1088/1757-899X/122/1/012036

Modeling of microporous silicon betaelectric converter with ^{63}Ni plating in GEANT4 toolkit*

P V Zelenkov, V G Sidorov, E T Lelekov, A Y Khoroshko, S V Bogdanov and A T Lelekov

Reshetnev Siberian State Aerospace University,
31, Krasnoyarsky Rabochy av., Krasnoyarsk 660037, Russia

E-mail: a.t.lelekov@yandex.ru

Abstract. The model of electron-hole pairs generation rate distribution in semiconductor is needed to optimize the parameters of microporous silicon betaelectric converter, which uses ^{63}Ni isotope radiation. By using Monte-Carlo methods of GEANT4 software with ultra-low energy electron physics models this distribution in silicon was calculated and approximated with exponential function. Optimal pore configuration was estimated.

Introduction

New generation micropower devices need compact and long-lifetime power supplies. One of perspective ways to create such devices is using Direct Conversion Nuclear Batteries based on the betavoltaic effect. Semiconductor betavoltaic converter of ^{63}Ni isotope radiation investigated by multiple scientific researches. ^{63}Ni isotope emits only beta-particles, without gamma and alpha radiation. Irridiation energy level below structural damage energy for most semiconductors and isotope half-decay time at about 100 years.

Isotope undergoes radioactive decay

$$^{63}_{28}\mathrm{Ni} \rightarrow {}^{63}_{29}\mathrm{Cu} + \mathrm{e}^{-},$$

the result of which is emission of electron (β-particle) having average energy 17.3 keV with a random direction. During travel through the space-charge region of p-i-n junction such particle causes the generation of electron-hole pair (EHP) which is being separated by the built-in electric field. Separated charge carriers travel to the opposite directions: holes into the p-region, electrons into the n-region. As the result, difference of potentials is induced on p and n region leads, the same way as it occurs inside photoelectric converter. If connectors are attached to external resistance, the current flows through the circuit (e.g. micropover voltage regulator) and generated energy can be used to power the load.

The main advantages of radioisotope-based power sources are long service life (determined by half-decay time), high specific energy, high energy density, wide range of operational environment conditions (determined by semiconductor) and high reliability. It is possible to create flexible

*Research is performed with financial support from Russian Federal Targeting Program "Research and development for priority targets of Russian scientific-technological complex development years 2014-2020", in accordance with the agreement 14.577.21.0117, unique identifier RFMEFI57714X0117

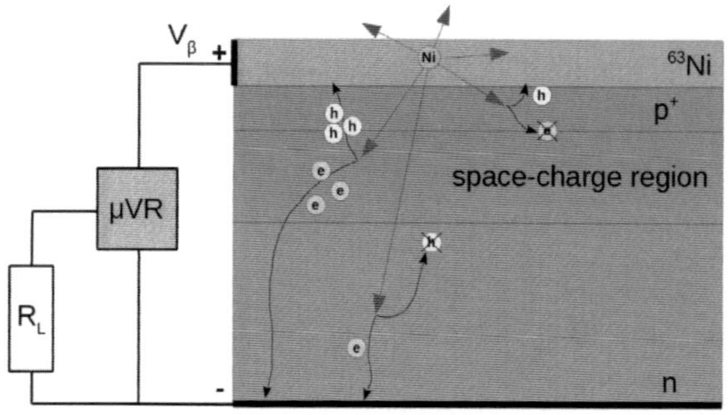

Figure 1. Direct Conversion beta-electrical converter principle of operation

hybrid voltage or current source incorporating the charge accumulation for continuous or pulsing mode operation.

One of viable designs of betaelectrical converter (BEC) is silicon wafer with deep micropores, every pore of which is turned into p-i-n junction and plated with ^{63}Ni. The main advantage of this design is significant increase of junction area and as the result – the increase of volumetric density of energy.

It is necessary to perform the modelling of all energetic processes – primary beta-electrons generation, electron-hole pairs generation, particles diffusion and drift inside space-charge region, recombination and others. It is possible to use COMSOL Multiphysics or Centaurus TCAD for modelling some of listed processes, but they do not incorporate the model of betavoltaic particles generation. For this reason Monte-Carlo method modelling using GEANT4 software was performed to estimate the distribution of EHP generation rate in semiconductor.

Construction definition

There are different possible designs of BEC: planar, microporous and microchannel with different geometrical parameters, microgranular designs are currently under development in modern scientific papers. The main aim of a switch from flat design to more advanced is to increase the junction area, because ^{63}Ni isotope has limited specific surface activity as deep placed atoms' radiation is shielded by surface atoms. ^{63}Ni is completely artificial, thus its cost significantly exceeds the price of silicon BEC - it is necessary to work out the design having optimal amount of this isotope. As was found out in earlier research [4], the total EHP generation curve as a function of isotope thickness is approximated by the exponent $a \cdot \exp(-\Delta y/T_y)$, where $a = 0,1878$, $T_y = 0,7987\,\mu m$. The layer having thickness of $3 \cdot T_y = 2,4\,\mu m$ generates 95% of useful energy, further increment of thickness does not increase useful energy generation.

Isotope application onto microporous silicon can be performed in different ways, one of the most simple is electrochemical deposition. When this method is used, nickel is deposed not only onto the surface of the pores, but also onto the outer surface of converter. As electrons are emitted by the isotope into a random direction, half of outer surface sourced radiation would be lost. To capture this radiation, extra BEC can be used, microchannel or planar type. When microchannel one is used, its outer nickel layer would screen some part of bottom BEC's radiation. Thus it becomes practical to use the planar BEC without isotope layer. In this case the design becomes hybrid as shown on figure 3(a), radiation of isotope is utilized in optimum way.

It is necessary to predict the optimum shape of microchannels on the basis of isotope surface layer area maximization and minimization of its volume.

The picture 2(a) shows the schematics of the oval microchannel, here R - radius of circular pore section, L - length of straight section, d - isotope layer thickness. Outer perimeter is estimated as $L_{\text{perimeter}} = 2\pi R + 2L$, isotope surface area as $S_{\text{Nilayer}} = \pi R^2 - \pi(R-d)^2 + 2Ld$. The isotope utilization efficiency is estimated as pore perimeter divided by isotope layer area:

$$k_{\text{eff}} = \frac{L_{\text{perimeter}}}{S_{\text{Ni layer}}} = \frac{2\pi R + 2L}{\pi R^2 - \pi(R-d)^2 + 2Ld}.$$

Let us assume the isotope thickness $d = 0.7\mu m$ and variate the values R and L in range, acceptable within current technology for microchannel manufacturing. Calculation results $k_{\text{eff}}(R, L)$ are shown on figure 2(b). One can see, that the extremal point lays near the minimum values of R and L. It is possible to conclude, that the most efficient shape is cylindrical pores.

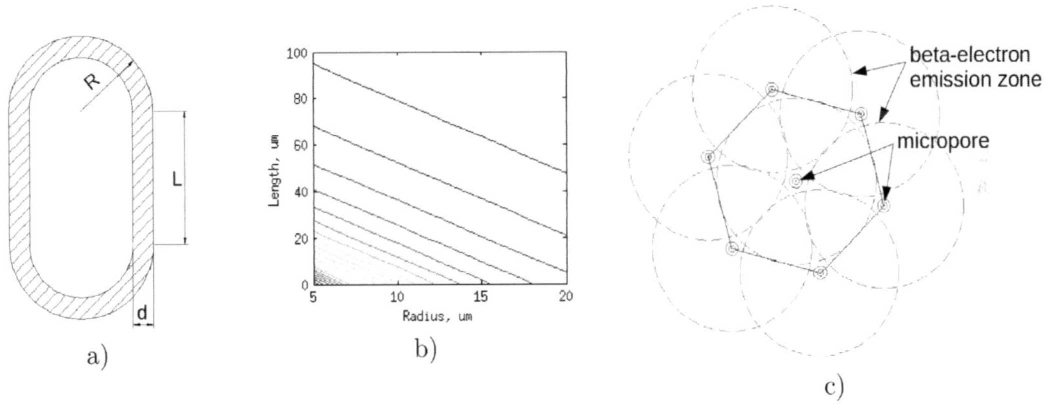

Figure 2. a) micropore geometry; b) plot of $k_{\text{eff}}(R, L)$ function; c) cellular micropore structure.

Circular microchannels should be placed on the surface of silicon wafer due to criteria of maximum surface utilization. As the pore is symmetrical along its longitudinal axis, the zone of carrier generation along the pore would be symmetrical as well. The maximum surface would be achieved when honeycomb placement is used, as shown on figure 2(c).

Computational problem definition

Current article propose to use the electron-matter interaction model, defined for low levels of energy. For instance, model proposed in [2] describe the ionization and electron transition into conduction zone for energy range from 16.6 eV to 50 keV. The calculated interaction cross-sections are used to plot the tracks of beta-electrons for isotope ^{63}Ni in silicon.

The selected geometrical schematic of this problem is shown on figure 3(a). The object is a cuboid, consisting of several parts: two silicon beta-electrical converter (BEC) of planar and micropore types, coated with layer of isotope Ni63, and air gap between BECs. Air also fills free space inside pore. Size of "World" region is $120\mu m$ in x and y dimensions. Planar BEC have thickness of $40\mu m$, micropore BEC – $140\mu m$. Pore diameter is $20\mu m$, depth is $100\mu m$. In conformity with article [1], economically efficient width of isotope layer taken $0.7\mu m$.

Fast electrons generation takes place in the isotope layer. The spectrum of beta-radiation corresponds to ^{63}Ni isotope (from [1]), approximated with exponential function $a\exp(-b \cdot E)$, where $b = 50.09\,MeV^{-1}$ (from [4]).

Particle tracks start from the isotope layer. Figure 3(b) shows the resulting beta-electrons tracks in silicon in cylindrical coordinate system (r, φ, z), averaged along the axis φ.

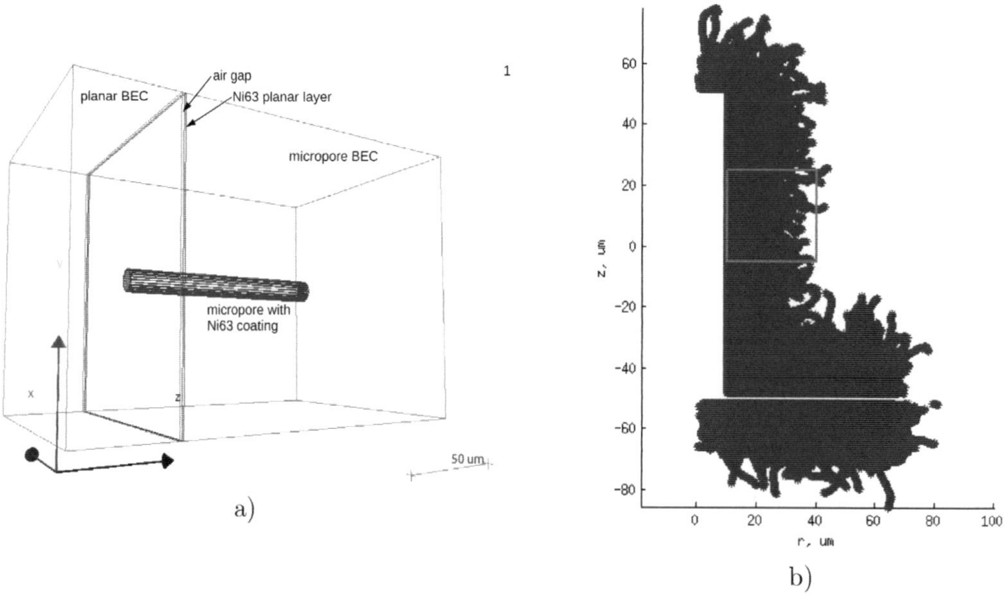

Figure 3. a) geometry of the problem; b) the tracks of beta-electrons, emitted by ^{63}Ni isotope (40000 tracks, cylindrical coordinate system).

Calculation sequence:

(i) Tracks calculation in GEANT4, corresponding to the software-defined experiment schematic;

(ii) Saving tracks data (nodal points coordinates, the amount and type of energy loss) as CSV-files;

(iii) Reading of generated CSV-files into memory, saving to float data files;

(iv) Separation the data of every track from whole data array, taking in consideration only ionization losses; secondary electrons are filtered out from tracks data, they are taken into account as separate particle tracks);

(v) Ionization energy losses (the number of electron-hole pairs generated) is calculated for every elementary volume of silicon.

(vi) Calculation of electron-hole pairs generation rate field in semiconductor.
The most informative in terms of p-i-n junction parameters optimization is middle, base part of micropore marked with red rectangle at 3(b).

Results and discussion

Representative segment of field was separated, and estimated field of energy generation rate was normalized relative to total initial energy of beta-radiation:

$$E_{\text{norm}} = \frac{1}{E_{\text{total}} \dfrac{S_{\text{pore side area}}}{S_{\text{total Ni63 layer}}}} \sum_{z} \sum_{\varphi} E(r, \varphi, z). \tag{1}$$

where E_{total} – total radiated energy, eV; $S_{\text{pore side area}}$ – pore surface area, taking part in the calculation; $S_{\text{total Ni63 layer}}$ – total isotope layer area; r, φ, z – coordinates of cylindrical coordinate system.

Normalized field of energy generation rate (one-dimensional normal to surface, pointed into the depth of semiconductor) was approximated using exponential function [5], translated at4.

Figures 4 and 3(b) represent the modelling results. The main modelling conclusions are: the betaelectron flow increase takes place near the corners, air gap of 1um does not shield betaelectron flow.

GEANT4 source code used for track estimation is uploaded to GitHub*. Number of tracks used for simulation was 40000, to increase the simulation precision. Calculation was separated into 4 threads to minimize the simulation time.

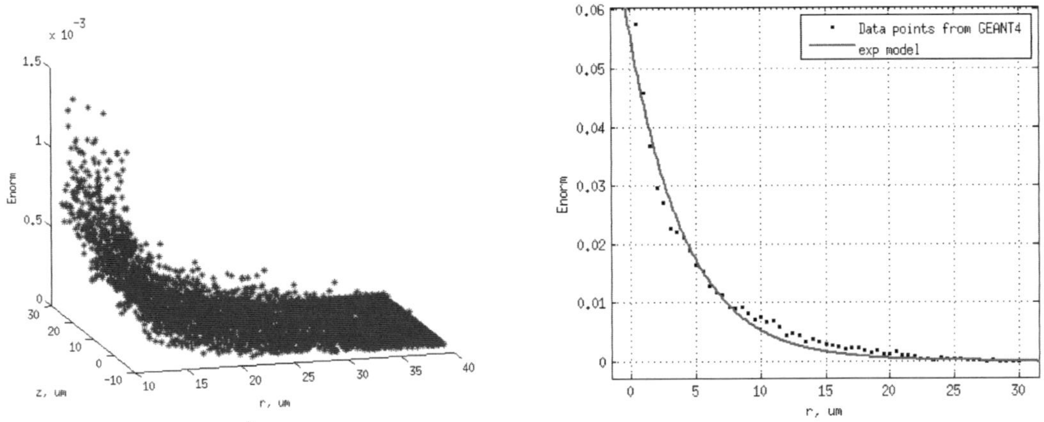

a) b)

Figure 4. Normalized energy field of carrier generation, a) 3D; b) 1D field.

Conclusion

As a result of the performed research it is revealed that

a) Minimum distance between micropores should be at least $20\mu m$. It is defined by maximum betaelectron energy. Maximum distance can be limited by the decline in silicon utilization efficiency. Optimum distance is more than $20\mu m$.

b) Micropore depth should be maximized and its diameter - minimized within limits of used technology for micropore creation and isotope plating. The shape should be as close to cylindrical as possible.

c) The highest generation rate density is near the open end of micropore. P-i-n junction quality should be the highest possible in this area to minimize parasitic currents.

d) The field of EHP generation rate was estimated, so that it is possible to optimize parameters of silicon semiconductor structure. Calculated exponential function approximation parameters can be used to define the fields for numeric modelling of beta-electrical converters, in software like Comsol Multiphisics.

References

[1] Loidl M, Le-Bret C, Rodrigues M and Mougeot X 2014 Beta spectrometry with metallic magnetic calorimeters *Proc. 19th Int. Conf. on Applied Radiation and Isotopes* **87** 302-305

[2] Valentin A, Raine M, Sauvestre J, Gaillardin M and Paillet P 2012 Geant4 physics processes for microdosimetry simulation: Very low energy electromagnetic models for electrons in silicon *Nuclear Instruments and Methods in Physics Research Section B: Beam Interactions with Materials and Atoms* **288** 66 - 73

*https://github.com/alex-khoroshko

[3] 2014 Geant4 User's Guide for Application Developers by Geant4 Collaboration. Version: geant4 10.1
[4] Kovalev I, Sidorov V, Zelenkov P, Khoroshko A and Lelekov A 2015 Electron-hole pairs generation rate estimation irradiated by isotope Nickel-63 in silicone using GEANT4 IOP Conference Series: Materials Science and Engineering **94** 012024
[5] Antognetti P, Antoniadi D, Dutton R and Oldham W 1982 *Process and Device Simulation for MOS-VLSI Circuits* (Urbino, Italy) p 619